French
in an Age

French Discovery in an Age of Revolution

World-Changing Scientific and Technological Advances, 1789–1815

JIM LIBBY

McFarland & Company, Inc., Publishers
Jefferson, North Carolina

ISBN (print) 978-1-4766-9213-5
ISBN (ebook) 978-1-4766-5015-9

LIBRARY OF CONGRESS AND BRITISH LIBRARY
CATALOGUING DATA ARE AVAILABLE

Library of Congress Control Number 2023026887

Front cover: Montgolfière balloon first manned ascent,
November 21, 1783 (© Archivist/Shutterstock)

Printed in the United States of America

*McFarland & Company, Inc., Publishers
Box 611, Jefferson, North Carolina 28640
www.mcfarlandpub.com*

For Donna

Table of Contents

Notes on the Text

Irregular capitalizations and misspellings of writings or quotes of the time may have been corrected herein without notation.

The English system of measurement is used. Prior to the revolution, France's measurement system varied by region. This was before the metric system was established in Revolutionary France in the 1790s. While the English system was not used in France, then or now, it will make things clearer to use a system that was not involved in the changing measures of France.

Words or phrases written in French will stand alone when the meaning is clear. If not clear, an English translation is provided.

Generally, kings and queens are referred to by their first names (Queen Elizabeth, King George) while military leaders are referred to by their last names (General Washington, Admiral Nelson). So, for the most part, Napoléon Bonaparte will be known as Bonaparte in his early years when he led troops and as Napoléon after he became the leader of France.

Preface

The inspiration for this book took place years ago. I was flipping through a high school chemistry textbook. Page one stated that Antoine Lavoisier was known as the Father of Chemistry and briefly explained why he was considered so. I had not heard of Lavoisier until that time. It was much later that I found out he died by guillotine during the height of the French Revolution. I did not see that coming! Scientists tend to lead important, but fairly mundane, lives. It might be something like finding out that Albert Einstein met his demise in the electric chair. How could society allow such a thing to even take place?

I later discovered other events of that time. Joseph Montgolfier was a contemporary of Lavoisier, born exactly three years before him. Montgolfier, joined later by his brother, began experiments in which they filled bags with heated air, larger and larger, until they were able to lift humans above the earth. I also learned how important this was. It was not simply a novelty, marking time until powered flight could be launched by another pair of brothers a hundred years later. It was certainly a novelty, but there were important applications to science, war, and even the entertainment field.

I knew the metric system was developed during the French Revolution, but that was about the extent of my knowledge. Various systems of measurement have been used throughout history and have typically evolved over the course of many generations. Somehow, the French, in the midst of revolution, developed a system in approximately four years that remains the standard through most of the world.

As time went by, I came across more information regarding this time. Eventually I decided I wanted to tell this story. I saw it as an astonishing time, not only because it seemed an amazing happenstance that so much discovery took place in one country at one time, but also that it happened in the worst of circumstances—in a country that was in constant disruptive change.

Introduction

Perhaps no object exemplifies the contradictions within France's revolutionary era as well as the guillotine.

In addition to being a well-respected physician, Joseph-Ignace Guillotin worked for medical reform in France, including the abolition of the death penalty.[1] When it became clear to him that the death penalty was not going away anytime soon, he sought to have it be made more humane. Throughout the world, capital punishment had a history of being cruel and painful. It was thought that making it so might discourage lawlessness. Hanging, burning at the stake, boiling, crucifixion, and a host of other methods were used. Those deaths were made public spectacles to serve as a greater deterrence, while also serving as a gruesome form of entertainment for the masses.

In France, beheading, done by axe or sword, was a right reserved for the nobility. Lower classes of people would typically be executed by a more drawn out and more painful method. However, even beheadings could be an agonizing form of death. Because it was a hideous task, it was often assigned to someone, perhaps a fellow convict, who was unskilled and perhaps plied with alcohol. If the executioner were not strong or accurate, it might take several attempts to achieve his goal.

In October 1789, less than three months after the storming of the Bastille, Dr. Guillotin recommended to France's legislative body that if executions were to be done, they should be done in a more civilized way. All classes of people should have the same method of execution. This method should be carried out by an impersonal machine and should be as quick and painless as possible. Guillotines were tall, so the blade would gain speed as it fell. The blades themselves were heavy and angled. All was done to result in a decisive cut. Though he neither designed nor built the guillotine, his name was forever attached to the device—something that horrified him for the rest of his life. His attempt to improve society became a symbol of the outrages of the French Revolution. An attempt to be more humane became a mark of inhumanity.

The chapters of this book outline the astounding discoveries that took place during this time. The amazing thing is that almost all these events took place during an era of revolution. The first chapter sets the stage by giving an overview of the French Revolution, especially topics that tie to later parts of the book. As with the chapter on the French Revolution, Napoléon Bonaparte is given a separate chapter to provide background. He was very much a direct and indirect part of the advancements of the age. From there, roughly chronologically, the innovations of the age are given separate chapters. Even in those chapters, there is a blending of scientific and political events. The final chapter shows several additional discoveries that perhaps do not warrant an entire chapter in length but are no less important.

The reader may notice that males played a prominent role in the events of this book. That is simply how the world was at that time. It was a male-dominated time, certainly in terms of science or technology. There are, however, women who were important through this story—the aeronauts Élisabeth Thible, Sophie Blanchard, and Elisa and Jeanne Garnerin. Marie Lavoisier was not as heralded as her husband, but likely his equal in intelligence and creativity. Without the intervention of Barbe-Térèse Méchain, the survey establishing the metric system might not have been completed. Other women, including Ada Lovelace and Hedwig Kiesler, play important roles.

The French Revolution, quickly followed by the Age of Napoléon, extended roughly from 1789 to 1815, though the seeds of the revolution had been growing throughout the previous decade. In *A Tale of Two Cities,* Charles Dickens wrote regarding this time and place, that "It was the best of times, it was the worst of times, it was the age of wisdom, it was the age of foolishness, it was the epoch of belief, it was the epoch of incredulity, it was the season of light, it was the season of darkness, it was the spring of hope, it was the winter of despair."

It was the worst of times—few countries, and few eras, could compete with the turmoil that existed in France at that time. The leader of the country, along with his wife, were arrested, then publicly executed as crowds cheered. Thousands of other citizens were executed by the government in the same fashion. There were riots in the country. It was difficult to keep track of all the factions within France competing for political power. There were Jacobins, Girondins, Hébertists, Feuillants, Montagnards, Cordeliers, and more. New constitutions were established in 1791, 1793, 1795, 1799, and 1804. The basic form of government dramatically shifted, beginning with an absolute monarchy, then a constitutional monarchy, then to various forms of a republic, an empire, a return to monarchy, back to an empire, and then back to a monarchy. As a republic, the governing body evolved from the National Assembly to the National Convention Assembly, the Legislative Assembly, the Directory, and finally a consulship. While this was taking place, France was in a fairly constant state of war beyond its borders throughout the 25 years. Possibly no country has ever had so much chaos in such a short span of time.

And yet, in many ways, it was the best of times. It is difficult to find a time in history in which there were more discoveries and inventions. This era saw the first automobile, human flight, the internal combustion engine, the parachute, and telecommunication. There was important, ground-breaking work in the theory of evolution, food preservation, and Egyptology. A legal code was developed that is still the rule of law in much of today's world. A method was developed to program looms which was later used to program the first computers. Modern chemistry, along with a new language of chemistry, was developed during this time. The secrets of hieroglyphics were unlocked. The metric system came into being.

It is amazing that such creativity and advancement took place in a country in utter turmoil. That is the story of this book.

I

Setting the Stage— The French Revolution

Louis Capet was in his cell in the Temple Prison. Capet was a family name, but not long before, he was Louis XVI, the King of France. His wife and two children were in the same prison, but he had been separated from his family by the authorities. His trial had just concluded. Louis had been accused of aiding France's enemies and of causing the hunger that plagued the country. For some it was not what he did, but who he was. Louis was a king and France was done with kings, even though the tradition of rule by monarchy was a thousand years old.

Louis was tried by France's current governing body, the National Convention. A total of 691 voted him guilty.[1] A number abstained, but no one voted for his acquittal. Now, what should be done with the former king? He could be kept in prison indefinitely. He could be banished. Thomas Paine, recently of America, was now living in France and had been elected to the National Convention. He asked that Louis be sent to the United States where he would be appreciated for the money and soldiers he had sent to America in their War of Independence.[2] However, the decision of the Convention was that he be executed. This vote was much closer, with just over half voting against him.[3]

He was allowed to visit his family the evening before his execution. His daughter thought his appearance was greatly changed since she last saw him. He and his wife Marie said their goodbyes. He spoke with his son and told him to forgive those that had rendered this verdict. He promised to visit them again the next morning but would not be able to make himself do so.[4]

That next morning, January 21, 1793, he woke at five o'clock and attended mass.[5] It was cloudy with a light rain as he got into the carriage that took him to the large public square, recently renamed the Place de la Révolution. There was a huge crowd, but also a large military presence to maintain order. Upon arrival, his coat was removed, his hands tied behind his back, and his hair cut to give access to the neck area. He was walked up the steps and laid face down on the plank at the base of the guillotine. The blade came down and the crowd cheered. They cheered again as his head was lifted by his hair and shown to them. Many kept as souvenirs the handkerchiefs they had dipped into his pooled blood.[6]

His wife, Marie Antoinette, now referred to as the Widow Capet, was soon separated from her children, first her son, and later, her daughter. The son, Louis-Charles, died in prison two years later, probably from tuberculosis likely aided by neglect and the unhealthy conditions in the prison. The daughter, Marie-Louise, was the only one of her parents' four children to survive to adulthood. She would later gain her release when she was sent to Austria, her mother's home country, in exchange for French soldiers imprisoned there.[7] She lived until 1851.

The Widow Capet was transferred to the Conciergerie Prison where she spent her remaining days locked in a small, dark room. Her trial and execution came nine months after Louis' execution. In a two-day trial, she was charged with such offences as conspiring with France's enemies, financial transgressions, and even a charge of committing incest with her son.[8] Her fate was the same as her husband's. Instead of being transported to the guillotine in an enclosed carriage, though, she was ingloriously sent through the crowd in an open cart. Like Louis, her head would be displayed to the cheering onlookers.

How does this happen in a country that sought to live by enlightened principles? Louis' and Marie's deaths were only two of what would become thousands of public executions. Along with these events, there was rioting throughout France and almost constant warfare with a host of other countries. It is equally amazing that this chaos could take place at the same time and place as many scientific discoveries that would dramatically change the world.

The Enlightenment

The French Revolution sprang from the French Enlightenment. The Enlightenment was a period when the human race began to, as the word implies, come out of the darkness. People were not destined to live the lives they led in the past but could improve their situations. Immanuel Kant, Isaac Newton, Francis Bacon, John Locke, and Thomas Hobbes explored the worlds of economics, science, and philosophy.

The French Enlightenment was a part of that movement, though its focus tended more toward solving societal ills. French Enlightenment savants sought to improve the world and the plight of those that had been left behind. It struck a chord with those seeking a change in the *Ancien Régime*; a monarchical, feudal, class system that had been the culture of France for centuries.

Jean-Jacques Rousseau (1712–1778) was a leading voice. He wrote that so-called primitive people were more civilized than the rich, sophisticated ones that society had placed above them. People had an innate human goodness. In his *Social Contract* he wrote that people had inalienable rights, not ones that needed to be granted to them by someone else, such as a king. There he stated that "Man is born free, but everywhere he is in chains." He disagreed with the emphasis on rote learning and felt children should learn by following their natural interests—a concept that greatly influenced how children would be educated in the future.

François-Marie Arouet (1694–1778), more commonly known as Voltaire, was a controversial figure. He was a Deist, believing there was a supreme power that created the universe and made it run. However, this supreme power was not the type of being that could be troubled with a person's prayers. He believed in the capability of humanity, so if the deity did not have time for us, that was all right. People would be fine on their own. It was a view not common in his time, but one that gained in popularity and was held by many of the influential leaders during the years of the revolution. Like Rousseau, he was a proponent of rights of the people. Their writings had a great impact on Americans such as Thomas Jefferson as they sought their own set of rights across the ocean.

Montesquieu (1689–1755) was another major influence on America's founding fathers, and many other leaders, as they sought to shape future governments throughout the world. He argued against despotic rulers that were atop so many countries at the time. He wrote of the need for a separation of powers. He said that separate branches of government were necessary to handle executive, legislative and judicial duties along with a system of checks and balances between those branches.

French Enlightenment authors collected much of the world's knowledge into a single source called the *Encyclopédie*. The coming French age of discovery would have been severely hampered without the knowledge its writings supplied. It sought to contain everything that was then known about the world as well as where the world should be headed. Men such as Rousseau, Voltaire, Montesquieu, and hundreds of others contributed articles. The collecting and editing of those articles were done by Denis Diderot and Jean de Rond d'Alembert, who also wrote hundreds of its articles themselves. The *Encyclopédie* was, in scope and size, beyond any other collection that had come before. Over a 20-year span, Diderot and d'Alembert created a resource that contained 20 to 30 volumes and over tens of thousands of articles. Religion, politics, science, and technology were covered—there was little not covered. Some articles were strictly informational, and others were quite opinionated, taking stands not popular with all, leading to its being banned by the Catholic Church.

Rivals and Allies

Much of revolutionary France was shaped from within but was also shaped by its relationships with other countries, especially England and the United States. England and France were separated by a 20-mile channel and had been at war for much of the previous millennium. In 1066, the Norman Conquest saw France invade British soil, and England would later return the favor. There were wars fought over religion, wars fought for territory, the War of the Spanish Succession, the War of the Austrian Succession, the Hundred Years' War, and the Seven Years' War. Seven different coalitions of countries fought France between 1792 and 1814 and England played a role in each.

The Seven Years' War, known in America as the French and Indian War, caused a financial strain for France that ultimately led to its revolution. There was not only the direct cost of the war, but in the treaty ending the war, France lost many revenue-producing overseas possessions. Lost territories were Canada, most of its claim to land along the Mississippi River, claims to India, and islands in the Caribbean. (The war proved costly for England as well. The British reasoned that their cost incurred during the French and Indian War should be shared by their colonies. The taxing of various items such as tea and sugar led to the American Revolution.)

A few years after the Seven Years' War, France chose to support the American colonies in their revolution against England. Whether this was because enlightenment principles supported the concept of independence, or simply because it served as an opportunity to gain revenge on their longtime foe, it produced an additional financial strain France could ill afford. Ultimately, France was on the winning side, but to loan funds to the American colonies, France itself had to borrow money, only compounding its problems.

Relations with the United States were generally less troubled. Though on opposite sides during the French and Indian War, they soon found a common enemy in England. The colonies would not have achieved their independence without France's help. Pierre de Beaumarchais, who went on to write *The Marriage of Figaro* and *The Barber of Seville*, channeled early covert aid to the colonist cause.[9] (Beaumarchais would be arrested in 1792 for criticizing France' revolutionary government, but later released.)

Further aid came in the form of loans negotiated by Benjamin Franklin. France not only supplied financial help, but also sent French troops to America. The Marquis de Lafayette came to aid the cause before an alliance between the two countries had even been reached. The first troops were brought by Admiral d'Estaing, who commanded a fleet of ships and 4,000 men. (Fifteen years later, he would face the guillotine in France, in part because he testified on behalf of Marie Antoinette at her trial.[10]) The crucial Battle of Yorktown would not have been won without French assistance, with nearly as many French troops there as there were colonial. General Washington commanded the colonial troops and General Rochambeau (who, during the later French Revolution, was arrested and nearly guillotined) commanded the French troops.[11] Admiral de Grasse's French fleet in the Chesapeake Bay kept Cornwallis from being rescued by the British navy.

In 1793, France abolished slavery, though a decade later it was restored in its territories by Napoléon Bonaparte. He did so to supply manpower needed to harvest the lucrative sugar cane and coffee crops of the Caribbean. When a Haitian rebellion led by former slave Toussaint L'Ouverture defeated the French, Napoléon decided to give up on expanding his empire into North America and sold the vast Louisiana territory to the United States.

There were difficult diplomatic moments such as the so-called XYZ Affair and the ensuing Quasi-War, which nearly led to actual war. The ongoing naval conflict between England and France would impact the United States and lead to the War of 1812.

The *Ancien Régime*

France had long been characterized by three major levels of societal structure. The clergy was known as the first estate. The Edict of Nantes, in 1598, gave French Protestants, known as Huguenots, religious rights. However, in 1685, Louis XIV revoked the edict, making Catholicism the religion of France. Much of the country's education and care of the poor was through the church, especially overseen by the Jesuits, an order of the Catholic Church. The second estate was made up of the nobility. Members inherited their titles or simply bought them. Like the first estate, they paid little or no taxes. During the French Revolution, many of the clergy lost their positions, but many of the nobility lost their heads. By far the largest of the three was the third estate, which was, by definition, everyone not a part of the first or second estate. They were made up of two groups. One group was the poor peasants who often worked the land within a feudal system. The other group was made up of what was known as the bourgeoisie—a middle class that was composed of lawyers, doctors, and businessmen.

The country was financed on the backs of this third estate. An organization known as the General Farm collected money from road and canal tolls and from a variety of other taxes. Members bought into the General Farm to collect those taxes for the state and then were allowed to keep a portion for themselves. However, those tax collectors would pay a heavy price during the years of the Revolution. There were many taxes. The work of the church was financed by a 10 percent tax on the people known as the *dime*. The *taille* was a land tax. The *corvèe* was an obligation in which the peasants were forced to give unpaid labor for upkeep of the nation's roadways. The *gabelle* was a salt tax. Taxation was a leading cause of the French Revolution—not just the taxes themselves, but the perceived unfairness in who was taxed and who was not.

Some of France's financial difficulties were due simply to misfortune. The country went through a series of poor harvests. Especially damaging was the summer of 1788 when France experienced a drought, along with major storms which devastated crops. This was followed by an especially cold, harsh winter—one that saw the Seine River freeze over. Not all of France's problems was bad luck, however. France's involvement in a succession of wars and overspending by royalty was certainly avoidable.

When Louis XV, the king of France, passed away in 1774, his grandson became Louis XVI. Those who knew him found the new king to be a decent, pious man. While a good man, he was probably a better person than king—one not always resolute in his decisions and socially somewhat inept. Though only 19 years old, he had already been married to Marie Antoinette for four years. Marie was from Austria, the daughter of Emperor Francis I and Maria Theresa. It was common for royal marriages to be arranged to firm up relations between countries. This particular marriage would have its difficulties in the future when their respective countries went to war and Marie's loyalty was questioned by the people of France.

A series of finance ministers, or comptrollers, attempted to find solutions to the country's fiscal woes, but little seemed to work. Finding a clear plan proved difficult as these advisors held different opinions concerning the wisdom of going to war, whether to loan money, whether to take out loans themselves, whether to increase taxes, or whether to make the tax structure more equitable. There were also agencies, though rarely used, that could be consulted during times of difficulty. France was a monarchy and the king ruled, but his ministers and these other agencies could serve in an advisory role. The Assembly of Notables was a collection of 144 individuals, mostly nobles, which had last been employed in 1626. They gathered at the seat of government, the Palace of Versailles, where they debated options, but with no consensus forthcoming. During this unhelpful process, Lafayette wrote to Thomas Jefferson asking if perhaps they should be called the "Not ables."[12] All that really was decided was to turn the matter over to another long-unused group, the Estates General.

The Estates General had been inactive for an even longer period than the Assembly of Notables, their last meeting having been in 1614. They gathered at Versailles in May of 1789. Each of the three estates were to send representatives along with a list of grievances which served as the basis of their work. Even when they met years before, their governing procedures had varied, so there was confusion as to how they would operate now. It was decided that in fairness to the third estate, being by far the most populous, that group would be allowed twice as many representatives. This concession to them was useless as each estate was given a single vote on the issues.[13] As a result, the third estate could still be outvoted by the other two.

The third estate came to one of the meetings only to find the doors to their meeting room locked. It is unclear whether this was an attempt to intimidate or a simple mistake. To those representatives, however, it seemed an obvious affront to them. Seeking another place to meet, they were led by Dr. Guillotin (he of everlasting infamy) to an indoor tennis court located on the palace grounds.[14] There they swore not to disperse until France had a new constitution and a new form of government. This became known as the Tennis Court Oath.

Some of the clergy began to splinter off and join with the third estate. More and more joined the group, and even though the king was not at all in favor of this turn of events, he finally simply admitted defeat and stated that all were to join this new body. It called itself the National Assembly. The empowerment of the third estate had a great deal to do with Emmanuel Sieyès, who had published writings in their support. Sieyès, would play an important role through the revolutionary years, culminating in a coup d'état that would bring Napoléon Bonaparte to power.

Louis did not handle these times particularly well. He went through a series of comptrollers looking for answers to the fiscal crisis. Jacques Necker, for one, was appointed to the position on three different occasions. Louis was initially firm in his opposition to this newly created National Assembly, only to cave days later. In part, he simply was not a firm decision maker, but he was also facing a personal tragedy. His son, heir to the throne, had just died from tuberculosis and Louis was deep in mourning.

The people of France grew increasingly short of food and patience, and their restlessness took the form of rioting. The Réveillon factory was a Paris papermaking plant. It saw days of violent demonstrations based on false rumors that wages were about to be cut. The Invalides, built as a home for disabled veterans, was attacked by a mob searching for weapons and ammunition. The Invalides attack took place on July 14, 1789. Later that same day, the mob moved on to the Bastille, a former fort that was currently serving as a prison, having housed luminaries such as Voltaire and the Marquis de Sade.[15] This day it held only seven people. Not only might the Bastille contain more weapons and ammunition, but it also was a symbol of the monarchy's authority over France. Bernard de Launay, who was in charge of the Bastille, must have felt well-protected behind walls that were five feet thick, eighty feet tall, and surrounded by a dry moat. However, he knew he only had limited supplies and could not hold out for too long. He thought the prudent thing to do was to surrender. He lowered the drawbridge and let the people in, which he probably thought would lead to his imprisonment. Instead, he and his men were promptly killed by the mob, with his and a Jacques de Flesselles' heads put on pikes and paraded through the city.

Louis was still king, but major societal changes were taking place which were well beyond his control. The National Assembly decided there would no longer be three estates. All classes and all people were equal. The feudal system was ended. Titles of nobility, such baron, or marquis, were gone. Regardless of background, a person was referred to as "citizen."

Enlightenment thought supported freedom of religion, so the church would not be abolished, though it would be greatly changed. Catholicism would no longer be recognized as the state religion. The Civil Constitution of the Clergy reduced the role of the pope, limiting him to giving guidance on spiritual matters, while the government oversaw the operation of the church. The number of clergymen was reduced, and those remaining were now employed by the state. Church lands were confiscated, with funds from their resale going to pay church salaries and to pay down France's debt. Clergy members would have to swear allegiance to the state to remain in their positions.[16] Many did so, but many refused.

Voltaire had written that "If God did not exist, it would be necessary to invent one." So, Maximilien Robespierre invented one, creating a Cult of the Supreme Being. A Festival of Reason was held at Notre Dame, though it, and other church buildings, were now renamed Temples of Reason. Names of months and days of the week were changed to ones that had no religious connotation. "Year One" of the new calendar did not mark the birth of Christ, but the birth of the new French republic.

The Fall

Louis had lost much of his authority, not allowed to do much more than simply rubberstamp actions taken by the legislature. He was given a suspensive veto allowing him to cancel a legislative action, and even those vetoes lasting only for a specific period of time. The National Convention Assembly (which had been the National Assembly for a period of three weeks and would become the Legislative Assembly two years later) was, in essence, now running the country.

In October 1789, more than 6,000 women marched from Paris to the Palace of Versailles, ten miles away, upset about the lack of bread. Lafayette, in charge of the National Guard, followed the marchers to attempt to keep the peace. At the Palace, to limit the violence, an agreement was reached that the government be moved to Paris where its actions could be more closely monitored. Louis and family (as well

King Louis XVI. Louis inherited a broken economy and years of bad harvests. He was a better man than he was a king. He lacked the ability to keep France from revolution.

as the decapitated heads of some of their guards) were escorted to Paris where they took up residence in the Tuileries.[17] From that time to the present, the French seat of government would be in Paris.

Louis was effectively powerless and now under something just short of house arrest. He made a plan to escape. Disguised and under the cover of darkness, he, Marie, and his children were taken by carriage, headed for Austria, Marie Antoinette's homeland. However, at some point, Louis was recognized, authorities were alerted, and a chase was on. Louis almost made it. Their carriage was halted in Varennes, only a few miles from the Austrian border. They were brought back, again placed in the Tuileries, but now with virtually no freedom or authority.

The situation was deteriorating. In the coming months, the Tuileries was invaded by mobs of citizens. Hundreds of Swiss Guards, there to protect the royal family, were killed. With a new constitution, the monarchy was formally ended, Louis was arrested, and his family was moved to the Temple Prison.

The Marquis de Lafayette had been the hero of two continents, but he could see his standing with the government steadily falling. He was a former noble. He was the head of the National Guard, charged with keeping order, which was the last thing some wanted. Lafayette was blamed for the escape of Louis, and was criticized when his National Guard fired on rioters at a celebration on the Champs de Mars. He considered himself a moderate, in favor of a constitutional monarchy, so now was branded by many radicals as a traitor. He left France when a warrant was issued for his arrest. Lafayette went to Austria, though this proved scarcely better since France and Austria were then at war, and being a French officer, he was arrested. His wife, Adrienne, worked tirelessly for his release, though she and her daughter came to be imprisoned as well, while several in Adrienne's birth family were executed in France.[18] He spent the next five years in foreign prisons until General Napoléon Bonaparte was able to negotiate his release.[19]

A coalition, including Marie's Austria, had formed against France. Those countries that were monarchies (which was most of Europe) were not at all happy with what France was doing with its monarch, and were afraid that there was a precedent being set which could spread to their countries. Toulon, an important port on the French Mediterranean coast, was taken by the British, but its later recapture was a first step in making a name for Napoléon Bonaparte. On France's eastern border, French troops were being pushed back. In 1792, in what was called the September Massacres, more than a thousand French prisoners were killed for fear that invading troops would liberate them so they could join the coalition forces.[20] However, a victory at Valmy turned the tide in favor of the French, though, in the aftermath, dozens of French officers were executed for what was considered poor performance on the battlefield.

Under a new constitution, a National Convention was formed to run the government. The conservative members, pro-monarchists, sat on the right side of the assembly hall and the radicals, revolutionaries, sat on the left. (The designation of "left" and "right" politics was born there and are still common political terms today.) During the first few years of the revolution, various political groups were formed and reformed as the leaders sought to lay their claim as to what direction the country should take. Their philosophies ranged from being somewhat different to being radically different from each other. One of the primary areas of conflict throughout those early years was what should be done with Louis XVI. What powers should he have as king? Should there be a king at all? Should he be tried for treason, and if convicted, should he live or die?

Ultimately, a trial was held, and Louis was overwhelmingly voted guilty of treason, with a slight majority then voting for his death. He was sent to the guillotine, as was Marie nine months later.

In the early years of the revolution, France had gone from an absolute to a constitutional monarchy. Then, under the Constitution of 1793 (which was referred to as the Constitution of Year I), France would begin its life as a republic. Legislation was handled by what was called the National Convention. The poorly named Committee of Public Safety served as an executive branch of the government. The first president of the committee was Georges Danton, but he lost power to the more radical Maximilien Robespierre, who became the de facto head of government. With France in danger of being overrun by foreign troops, the committee was given extreme powers which they fully employed. An enacted Law of Suspects authorized the use of tribunals and was used to arrest thousands. Treasonous activity (a charge that could be made to fit many situations), speaking out against the war, or even failing to address others as "citizen" could be cause for arrest. There were trials, but suspects were generally considered guilty until they could prove themselves innocent.

Later, the Law of 22[21] Prairial was enacted, which extended the reach of the tribunals. Citizens were now obligated to turn in those they suspected of wrongful activity. This law also ended a person's right to representation. With no lawyers to offer up a defense, the accused had to answer the charges the best they could. Courts simply reviewed the charges, then made one of two choices—death or acquittal. Referred to as the Reign of Terror, this period lasted less than a year, but there were more than a quarter of a million arrests and an estimated 17,000 executions.[22]

Following The Terror was simply a different kind of terror. The excesses of the Committee of Public Safety came full circle to what was known as the Thermidorean Reaction. This period saw those who were sending thousands to the guillotine now come under judgment themselves. Jean-Paul Marat was a leading voice on the Committee of Public Safety. One day while he was bathing (something he often did due to a persistent skin condition) a young woman, Charlotte Corday, came to his house and asked permission to meet with him. She was a member of the Girodins, a party that was in opposition to Marat's Jacobin party. Initially turned away, she was eventually allowed entrance to his home, and while he was still in his bathtub, stabbed him to death. She was tried and sent to the guillotine. The next year, Robespierre was arrested, and after a failed suicide attempt in prison, he was sent to the same guillotine he had sent so many to. Georges Danton's fate was the same.

There was a new constitution in August of 1795. As the United States government had established, there were now two legislative chambers,

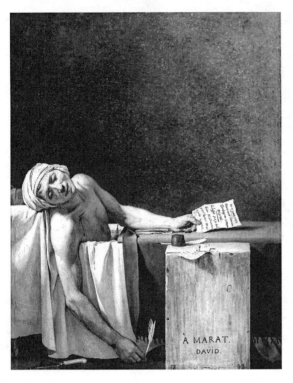

A MARAT.
DAVID.

The Death of Marat was painted by artist and revolutionary Jacques-Louis David. Marat was a Jacobin revolutionary. **He was murdered by Girondist Charlotte Corday, but in doing so, she made him a martyr. He played a key role in the death of Antoine Lavoisier.**

one consisted of a Council of 500, which proposed the laws, and a Council of Ancients which accepted or rejected those proposed laws. A five-person Directory acted as an executive branch, seeing that those laws were carried out.

In time, one of the Directory members, Emmanuel Sieyès, planned a takeover of the government. He felt he needed a military presence to join him to ensure his success, which he found in the person of Napoléon Bonaparte. The whole affair, which began as a bit of a disaster, was ultimately a successful coup. However, Sieyès would not have the power he thought he would. Bonaparte, Sieyès, and a Roger Ducos formed a three-person consulship, with Bonaparte soon taking preeminence. He began as First Consul holding a ten-year term, then First Consul for life, and finally, emperor. It was the beginning of his domination of the life of France and the beginning of the Napoléonic Age.

II

The Advent of Flight

"Surely the sky lies open, let us go that way."
—*Ovid's Metamorphoses*

"When the boy began to delight in his daring flight, and abandoning his guide, drawn by desire for the heavens, soared higher. His nearness to the devouring sun softened the fragrant wax that held the wings: and the wax melted: he flailed with bare arms, but losing his oar-like wings, could not ride the air. Even as his mouth was crying his father's name, it vanished into the dark blue sea, the Icarian Sea, called after him. The unhappy father, now no longer a father, shouted 'Icarus, Icarus where are you? Which way should I be looking, to see you?' 'Icarus,' he called again. Then he caught sight of the feathers on the waves, and cursed his inventions. He laid the body to rest, in a tomb, and the island was named Icaria after his buried child."
—*Metamorphoses Book VIII*

Mankind has always been fascinated by flight. In Greek mythology, Daedalus fashioned wings for his son Icarus and himself to escape from the island of Crete. A collection of Middle Eastern stories, *Arabian Nights,* tells of riding a magic carpet. Leonardo da Vinci was fascinated with the human potential for flight. From his observation of the flight of birds, he designed vehicles in which a man, by his own power, could fly. Regarding Leonardo's fascination with flight, Walter Isaacson states that "he produced more than five hundred drawings and thirty-five thousand words scattered over a dozen notebooks on these topics."[1] The Wright brothers worked for years to demonstrate that man could fly in a powered, heavier-than-air machine. However, the Wright brothers were not the first to fly. Their achievement came 120 years after brothers Joseph and Étienne Montgolfier launched the age of flight.

The Montgolfier family had been in the paper manufacturing business since at least the 1500s. They began in the provincial town of Annonay,

France, eventually expanding the business to include many other paper mills. The quality of their papermaking led Louis XVI to declare their business a "royal manufactory."[2] Although under a different name today, it is still making paper, with its products sold worldwide. Their hometown and site of their first factory, Annonay, would see the first public launching of their balloon.

The situation was suited for Joseph and Étienne to explore flight. The success of the family business gave them the time, the finances, and the material resources necessary to devote to the task. The brothers also received a solid education, which was not common in those days. Joseph attended school in their hometown of Annonay and then a Jesuit school in Tournon. Étienne, the younger, and seemingly the more studious of the two, attended school at the Collège Sainte-Barbe in Paris. Both were educated in the sciences—mechanics, architecture, math, chemistry, and statics—all areas that would prove valuable in their future efforts.

Joseph was born on August 26, 1740, and Étienne on January 6, 1745, to parents Pierre and Anne. Étienne, although the second youngest of 16 children, was chosen by Pierre to be in charge of the family factory after the eldest son passed away. This prematurely ended his schooling in Paris. He was 27 at the time and spent the next ten years developing and improving the factory and the papermaking process.

The business of papermaking had a direct application to their first flights. A lightweight material was needed for the envelope. Paper was light and they were experts in its production. Early trials were made of paper and often layers of paper. The historic first public flight in 1783 used an envelope whose construction was an outer covering of sackcloth lined with layers of paper. As their balloons increased in size, paper eventually would not be a strong enough material. Over time, paper would give way to rubber, nylon, and various fabrics and plastics.

Initial Experiments in Flight

It is unclear exactly what sparked Joseph Montgolfier's interest in flight, but he stated that an important step was when he read English chemist Joseph Priestley's *Experiments and Observations on Different Kinds of Air*. Priestley's book had recently been translated into French. That there were "different kinds of air" was a major change in thought for the science of that day. Since the time of Aristotle, the world was believed to be made up of basically four substances—earth, air, fire, and water. Science was on the verge of a general acceptance that air was not a single substance, but a mixture of several different elements and compounds.

As the smoke from a fire rose into the sky, it was clear that heated air rose, but it would turn out that this would not be the only option for flight. Henry Cavendish, an English chemist, managed to isolate one type of Priestley's "different kinds of air." He called this "inflammable air," later to be known as hydrogen. For now, it was called inflammable air because it was combustible (which would have tragic consequences in the future of flight). The importance of this discovery was that this different air, whatever it was, was much lighter than common air. Depending on such qualities as purity, temperature, and humidity, this inflammable air

Joseph Montgolfier ushered in the age of flight. He shared his initial experiments with his brother Étienne, and together they sent people into the skies for the first time in history.

seemed to weigh somewhere between one-tenth and one-fifteenth what common air weighed. An object filled with this air, being lighter, should rise through the surrounding atmosphere, just as Archimedes had stated two thousand years before when he wrote *On Floating Bodies*. His Proposition 5 states that "Any solid lighter than a fluid will, if placed in the fluid, be so far immersed that the weight of the solid will be equal to the weight of the fluid displaced." And in Proposition 6, he states, "If a solid lighter than a fluid be forcibly immersed in it, the solid will be driven upwards by a force equal to the difference between its weight and the weight of the fluid displaced."

If an object is placed in water and it is moving neither up nor down, that must be because it weighs the same as the volume of water that it is replacing. It is at a point of equilibrium. However, if an inflated beach ball is held under water and then released, it will rapidly move upward. It rises because whatever the ball weighs is less than what the same volume of water would have weighed. Archimedes probably had water primarily in mind when he wrote about fluids, but his propositions are true for all fluids, including gasses. The first flights would be a contest between heated air and hydrogen as to which was a better method to achieve lift.

Joseph Montgolfier was the first of the brothers to explore the possibility of flight. In 1780, he was in Avignon with his brother Alexandre, both training for a career in the law. It is there that Joseph began to experiment. For Joseph's initial experiments, he filled small paper bags with heated air and watched them rise to the ceiling. They rose, but the air inside soon cooled, and the bags drifted back down.

He continued to experiment. He thought that perhaps the heating of the air produced a new kind of gas that was lighter than the surrounding air. He was wrong about that. Heating the air was only a physical, not a chemical change. Though Joseph did not know it at the time, what took place was that as the air molecules were heated, they became more active and more spread out. The lesser density inside a container would then cause it to be lifted through the more dense, unheated air.

What Joseph did know is that the heated air worked. He theorized it might have something to do with the smoke from the fire. If smoke was the lifting agent, he figured the more of it, and the more the smell, the better. Thus, in the future, there would be a number of unconventional fuels used, including wet straw, wool, old shoes, and decomposing meat.

Joseph constructed a box that was three feet by three feet at the base and four feet tall, made of very thin wood strips and covered in a light taffeta cloth.[3] After lighting some paper and inserting it inside an opening at the bottom, the box rose to the ceiling and stayed there until finally descending. He wrote brother Étienne in November of 1782, "Get a supply of taffeta and of cordage, quickly, and you will see one of the most astonishing sights in the world."[4]

Reunited in Annonay, the brothers got to work. The work was not simply random guessing. A good deal of math and physics had to be done, such as calculating surface areas, volumes, and lifting forces. Most of that calculating work was done by Étienne.

Joseph showed Étienne his original design which ascended 70 feet on a test flight.[5] Their next balloon's envelope was built with almost ten times the capacity of the previous one.[6] Subsequent balloon flights would see larger balloons and greater heights achieved. These were meant to be tethered flights, but those flights did not always proceed according to plan. Cords can break or individuals holding the cords might accidentally let the balloon escape prematurely. On one occasion, one of their balloons escaped its tether and did not come down until rising 6,000 feet and traveling over a mile away.[7]

Just how large would these balloons need to be? A child's balloon can be relatively small because there is very little to lift. The envelope, likely made of latex, is heavier than air, but still, not very heavy. The total weight would be small, thus it would not take very much to achieve lift. (Helium is the gas inside a child's balloon today. It would be used rather than hydrogen because it is not flammable—a fact that could have saved the lives of many early balloonists. Unfortunately, helium was not discovered until 1868.) Joseph's paper bags, just like a child's latex balloon today, are extremely light and would not take very much lighter-than-air gas to achieve lift. However, these balloonists would not settle for lifting a thin sheet of paper. Soon, these flying machines would be used for scientific research, reconnaissance in times of war, and transportation, even though these applications might not have been envisioned at the time. The total weight would include the air inside the envelope, the envelope itself, a basket, any number of human beings, instruments, etc. These flying machines would soon weigh over a thousand pounds. Also, the envelope itself could not be as light a weight any longer. It would need to be sturdy enough to withstand hazards such as random tree limbs, rough landings, wind gusts, and sparks from attached fires. It would also need to be heavy enough to withstand the difference in pressure between the air inside and outside the envelope without bursting.

If an object was to get off the ground, its overall density would have to be less than that of the surrounding air. At sea level, with an outside temperature of 68 degrees Fahrenheit, air has a density of 0.0752 pounds per cubic foot. Air heated to a temperature of 200 degrees has a density of 0.0601 pounds per cubic foot. (Air much hotter than that would be less dense but would likely cause damage to the envelope.) However, the parts connected to the balloon are, comparatively, very dense. For example, the human body has a density of roughly 62 pounds per cubic foot. These balloons would soon take along bags of sand to serve as ballast. Sand has a density of almost 100 pounds per cubic foot. Additionally, the pilot might bring equipment made of lead, which has a density of 707 pounds per cubic foot. There would need to be a very large volume of this lighter air to compensate for all this extra weight. Future lighter-than-air flight would only be accomplished with an extremely large balloon attached. And as the payload grew, so must the size of the balloons.

The First Public Flight

Most of these early, experimental flights were only witnessed by family, but word was starting to get out. Joseph and Étienne felt it was time for a public demonstration. Paris would be the ideal location, but first, they wanted one more trial to make sure everything was ready. They held a public showing in their hometown of Annonay. On June 4, 1783, a mixed crowd of nobles and peasants gathered to view the event. The Montgolfiers had constructed their largest balloon yet. It was made from sackcloth and lined with three layers of paper. The balloon had three red and yellow bands running horizontally which were held together with just under two thousand buttons. The bands overlapped slightly so that one could be buttoned into the adjacent band, and they formed a balloon that was roughly spherical. In a letter, brother Alexandre stated the balloon weighed 500 pounds and had a diameter of 35 feet.[8]

The uninflated envelope was suspended upright, attached to poles on each side to hold it up as it was being inflated. This would be the common practice of filling for almost all the initial flights. A fire was kindled underneath its bottom opening, lit with the help of alcohol and fueled with wool and straw.

As it inflated, volunteers held the balloon in place. When inflation was complete, Étienne gave the command, and those holding onto its ropes released them, and the balloon went straight up into the air. It went over a mile into the sky, and stayed aloft for ten minutes, landing a mile and a half from where it started. There would soon be longer flights. This one likely came up short because, as Étienne wrote, "The loss of gas by the button-holes and other imperfections did not permit it to continue longer."[9] Word traveled quickly and news of their flight spread throughout the country. The brothers soon received invitations to demonstrate their device before the French Academy of Sciences and with the king and queen in attendance no less. Before this demonstration took place, however, other entrepreneurs were pursuing flight and with a completely different source of lift.

Hydrogen Balloon Flight

Jacques Charles was a 37-year-old physicist. He would be credited with Charles' Law, a principle which states that the volume and temperature of a gas are proportional. He was familiar with the discoveries regarding Joseph Priestley's inflammable air. He also knew of the flight of the Montgolfière balloon. He meant to duplicate that flight; however, he erroneously assumed the Montgolfiers were using inflammable air and not simply heated air.

He enlisted the aid of two brothers, Anne-Jean and Nicolas-Louis Robert, to construct a balloon. These Paris craftsmen designed an envelope made of fine silk with a rubber coating underneath. This would be a much smaller balloon that the Montgolfier's. Decked out in red and blue, it would be 13 feet across—well less than half of the one that flew in Annonay.[10] More important than the smaller diameter was that it contained approximately one-twentieth the air inside. This reduction in size was possible because of inflammable air's (hydrogen's) greater lifting capacity, being much less dense than the heated air used by the Montgolfiers. While air, at a temperature of 200 degrees, has a density of 0.0601 pounds per cubic foot, hydrogen has a density of only 0.0051 pounds per cubic foot. Even at this smaller size, though, generating enough hydrogen to fill the balloon would be a challenge. Hydrogen had been isolated, but only in laboratories and only in relatively small amounts. Quantities of hydrogen that were measured in so many cubic inches would now be needed in an amount of over a thousand cubic feet.

Chemists of the day knew that the needed hydrogen could be isolated by pouring sulfuric acid over iron filings, with hydrogen bubbling up from the reaction. The Robert brothers constructed a device that would siphon off the gas, carry it through pipes, and collect it in a larger main pipe, which then led to the balloon. A thousand pounds of iron filings and almost 500 pounds of sulfuric acid was used during the process.[11] This would be a challenge even if everything went well. But things did not go well. First, the device leaked and had to be rebuilt. After this rebuilt machine began working, the chemical reaction generated heat which, then, dangerously overheated the balloon fabric. The fabric had to be cooled by pouring water over it, but then water vapor, which had condensed and formed inside the balloon, had to be removed. That water was somewhat acidic, and, while inside the balloon, began to eat away at the fabric. It took four days before the balloon was finally filled and ready.

There was a tethered launch trial in the middle of Paris in which the balloon rose 100 feet in the air.[12] Predictably, the balloon brought a crowd of people to the launch site, and Charles decided a more open area would need to be found. To avoid the masses, it was moved in the middle of the night from its inflation site at the Place de Victoires, through the streets of Paris, to the more spacious Champs de Mars, a distance of about three miles.

In a few years, the Champs de Mars would see some of the best and worst of this revolutionary era. For many years it had been an open area in which Parisians farmed individual gardens. Its development to more of a park-like setting began in 1760 when its southern end became the home of the École Militaire. (The Eiffel Tower would sit at its northern end a hundred years later.) On July 14, 1790, it would be the setting of a giant celebration on the first anniversary of the storming of the Bastille.

In 1791, Louis XVI had just attempted his escape from France only to be caught and brought back to Paris. In a show of leniency, the National Assembly allowed Louis to remain king of a constitutional monarchy, though with curtailed powers. Many felt that was too lenient and assembled to let their voices be heard. In an escalating situation, the mayor of Paris, Jean-Sylvain Bailly, declared martial law. The Marquis de Lafayette, as commander of the National Guard, attempted to restore order. In a scene reminiscent of the Boston Massacre twenty years earlier, a crowd began to hurl rocks at the French troops who then responded by firing. This event came to be known as the Champs de Mars Massacre with dozens being killed. As mayor, Bailly would be held responsible.

Jean Bailly had been a respected astronomer before the events of the revolution. He had plotted the orbit of Halley's Comet, studied the moons of Jupiter, and wrote several books on astronomy. When the Estates General were called to meet, he became a key figure. Bailly was the leader of the third estate when it made its stand that resulted in the Tennis Court Oath. He was chosen to be the president of the National Assembly and then the mayor of Paris. As part of the fallout from the Champs de Mars Massacre, he resigned his position. He was arrested two years later and put on trial, accused of causing the massacre and as well as assisting the king's and queen's attempted escape. The first charge was dubious and the second simply wrong. Two days after the trial, Bailly went to face the guillotine at the same Champs de Mars.

But now, on August 27, 1783, the Champs de Mars would host the first ever hydrogen balloon flight. During the move from the Place de Victoires, some of the gas escaped, so more needed to be added. A huge crowd had been waiting throughout the day and was growing impatient. Finally, at five o'clock that evening, it was ready. A cannon shot announced the launch. The balloon went into the clouds, but it reappeared, which warranted another celebratory cannon shot.

Leonhard Euler, one of the world's greatest mathematicians, was nearing the end of his life. He had heard the news of the flight. Living in St. Petersburg, and blind, he worked out mathematical calculations regarding this balloon flight, demonstrating, as he said, "The laws of vertical motion of a globe rising in the calm air in consequence of the upward force owing to its lightness."[13] He would die three weeks later, but his son sent his calculations on to the French Academy of Sciences.

Mathematicians were not the only ones interested in the flight. The whole city watched. This, as well as many future flights, were financed by giving special, up-front seating to those willing to pay. For one crown, an individual and two guests could get special seating. As the old regime lost favor, the "crown" would soon be a thing of the past, but this practice of paying for prime seats would continue to be a way of financing flights.

Upcoming flights would also be funded by obtaining sponsorship by individuals, science academies, or other interested groups. For those that chose not to pay, there was still plenty of good seating available. People came out to the streets and the rooftops to see this amazing sight. Truly, anyplace that had an unobstructed view of the sky was a good seat.

One of the spectators for Charles' flight that day was Benjamin Franklin. He was in France to negotiate the end of the American colonists' War of Independence against England. The Treaty of Paris would be signed a week later. As the city viewed this balloon flying through the air, Franklin wrote, "Someone asked me—What's the use of a balloon? I replied—What's the use of a newborn baby."[14]

The voyage did not end well. After traveling 20 miles to the northwest, the envelope tore. This could have been due to the changing air pressure as it rose. As a balloon's altitude increases, the surrounding air becomes less dense, causing the balloon to expand, and this expansion could have caused a tear in the fabric. For later flights, that problem could be remedied by venting some of the interior gas, but this time it meant the end of the trip.

Other than the cities of Annonay and Paris, most did not know these huge flying objects existed, or that flight was even being considered. As the balloon floated over the French countryside, peasant farmers were tending to their fields. One can imagine their shock as they looked upward and witnessed this contraption in the air. And then when it crumpled and fell to the earth, those farmers had to have been traumatized. Taking no chances on what it might be, they decided to make sure it was dead. They attacked it with pitchforks. For good measure, one thrust a sword into the envelope resulting in a rapid release of the remaining hydrogen gas, which must have frightened them even more. As a finishing touch, they tied it to a horse and sent the horse galloping away.[15]

Any balloon flight was likely a shocking spectacle to anyone that saw it for the first time. It is difficult to imagine what a person must have thought on the first occasion of seeing an airplane, a television, or an automobile. As communication improved, people would have some idea what might be coming, but many in the late 1700s had no clue that a concept such as human flight was even being considered. As with the farmers' encounter with the Charles balloon, its appearance must have been frightening.

In an attempt to lessen the trauma, the French government issued the following, dated September 3, 1783: "The first experiment was made in Annonay in Vivarais, by [Monsieur] Montgolfier, the inventors; a globe formed of canvas and paper, 105 feet in circumference, filled with inflammable air [incorrect], reached an uncalculated height.... It is proposed to repeat these experiments on a larger scale. Anyone who shall see in the sky such a globe (which resembles '*la lune obscurie*' [the dark moon]), should be aware that, far from being an alarming phenomenon, it is only a machine, made of taffetas, or light canvas covered with paper that cannot possibly cause any harm and which will someday prove serviceable to the wants of society."[16]

Animals in the Air

The Montgolfiers' first flight took place in Annonay, a town of only a few thousand. Though it was almost 300 miles away, Paris was the most important city in France and was also home to the prestigious French Academy of Sciences. They doubtless also felt the need to match the recent Parisian flight of Jacques Charles. To make the connection to the Academy, Étienne enlisted the help of an acquaintance, Nicolas Desmarest.

Étienne had met Nicolas during his college years in Paris. Desmarest had previously made important discoveries regarding the earth's history. Basalt was known to be a common rock underlying the earth's surface. Thought to be a sedimentary form, Desmarest showed that it was in fact igneous, and its source was volcanic activity. It was an important discovery, revealing the role volcanoes played in the development of the earth.[17] In 1757, Desmarest was employed by the French government to create better manufacturing methods and received promotions within the royal government up to the time of the French Revolution. That, though, turned out to be a bad time to have any association with the royal government. Because of that association, he was imprisoned for a time, but did manage to survive and in fact lived to the age of 90.

Because of his work in geology, Desmarest was elected to be a member of the Royal Academy of Sciences. This gave him important contacts that would prove useful to Étienne. Established by Louis XIV the previous century, the Academy included the leading scientists of the day. For the Montgolfier brothers, in 1783, this was the organization to impress. Étienne wrote to Desmarest, "But so that no one poaches on our preserve in the meantime, I beg you to announce it in our name to the Academy, or if that cannot be done, in some journal, so that a fixed date may result."[18]

The flight was scheduled to take place in the courtyard of the palace of the king. The brothers were also invited to give a demonstration flight for representatives of the Academy of Sciences. The eight-person committee included his friend Nicolas Desmarest, along with luminaries such as Antoine Lavoisier and Marquis de Condorcet.

Joseph arrived in Paris to begin the work on his balloon and was there in time to witness the flight of Charles' hydrogen balloon. The committee said they would take care of the expenses involved, although that cost was ultimately taken on by the French government.

Étienne, Ami Argand, and Jean-Baptiste Réveillon directed the work. Argand and Réveillon were not only important in this endeavor but were key individuals in French society as well. Argand was a 33-year-old scientist. It was during this time period that he was working on what would become known as the Argand lamp. Although light was already being provided by candles and lamps, his work led to a lamp that was as much as ten times as bright as earlier models.[19] It became the most popular source of nighttime lighting until the 1850s and the advent of the much cheaper kerosene lamp. Commonly, before his lamp, when it grew dark in the evening, people simply went to bed. The Argand lamp's need for oil was a major reason for the growth of the whaling industry (as evidenced by stories such as Herman Melville's *Moby Dick*). He lost everything in the French Revolution and died in 1803 at the age of 53.

Like the Montgolfiers, Jean-Baptiste Réveillon was in the papermaking business. It was at his Paris factory that they would build their next balloon. Considered by some to be the first act of violence of the revolution, in 1789, a disturbance would take place at this very factory site. These disturbances were known as the Réveillon Riots and came three months before the storming of the Bastille. Rumors spread, unfairly, that Réveillon was planning to lower wages, at a time of food shortages following an extremely difficult winter. On the contrary, Réveillon had actually lobbied the government to lower food prices, but the talk among the employees was that he was working against them. The result was the destruction of the factory and dozens of deaths and injuries over the course of four days. He and his family escaped harm only by climbing over their garden wall as the mob descended on their home.

Prior to this, in 1783, Réveillon's factory was the site the Montgolfier brothers had chosen to assemble their balloon. As it was being constructed, the envelope pieces were laid out in the Réveillon factory courtyard. Having learned the lesson that attaching pieces using buttons led to air loss, the pieces were now sewn together. When finally assembled, it had "a very odd shape"[20]—the top and bottom seemed to be something like truncated cones with a boxlike middle section. The envelope was cloth, lined with paper. Joseph felt cloth with a coat of varnish would be strongest. Even though they did not have to worry about the cost, Étienne nixed that idea as being too expensive. He would come to regret that decision. They used cloth and decided to include a paper coating on each side of the cloth.

Earlier tethered trials had gone well, and all was ready for their demonstration before the Academy's committee. It would take place Friday morning, September 12. It only took ten minutes for fire to heat the air and inflate the balloon. However, timing of the flight coincided with ever-increasing rain which turned into solid rain and gusty winds. The flight was tethered because the plan was to reuse the balloon for the flight before the king. Argand felt that less damage would be done if the balloon was released and allowed to go with the wind rather being tied in place and battered. Argand was probably correct since the balloon did fly, however it took a beating doing so.

Ultimately, the flight had to be considered a success because the committee members saw what they came to see. The Montgolfier brothers had, in fact, achieved flight. The balloon, though, was in bad shape. The outside paper had turned into a soggy mess and the envelope was ripped to the point of being unusable. A week remained until the scheduled presentation before the king of France, and they had no balloon.

They got to work, and amazingly, completed the task of constructing a new balloon in four days. It was 57 feet tall and 41 feet across and had a more typical looking balloon shape than the previous one.[21] This time, Joseph's original thought of a cloth balloon covered with a coat of varnish was adopted. During a trial run, there was a tear along a seam, but it was quickly repaired.

They were ready. The uninflated balloon was transported the ten miles from Paris to the Chateau de Versailles. Versailles was the seat of government and home to the king and queen as it had been for over a hundred years, from the days of Louis XIV. Louis had the palace built and transferred the government there from Paris, with the huge expense of building and maintaining the palace being no help to the country's financial difficulties. It would be the destination of the Women's March in six years. After being escorted to Paris, neither Louis nor Marie nor any other French royalty would ever reside there again. In the days following Louis' departure, the art was taken to the Louvre and the furniture auctioned off.

In 1783, however, the palace was magnificent; likely as magnificent as the restored palace is today. On the Montgolfier's flight date, September 19, there was a party-like atmosphere at Versailles. The palace itself had thousands of residents on typical days. This day was not typical, though. Spectators included those that came from Paris and the surrounding countryside, members of the French Academy, and of course, King Louis XVI and Queen Marie Antoinette. The launch point for the flight would be the large courtyard in front of the palace.

In the middle of the courtyard, a raised, wooden, eight-sided stage was built that was tall enough that people could walk under it. In the middle of the stage was a large, circular opening which would host a four-foot-tall brazier which would be used to heat the air.[22] The king and queen were given a close-up tour of the layout. When the fire was lit, with its typical Montgolfian odor, the royal couple retreated to the palace. To add to the drama of the event, the Montgolfiers decided to send living beings on the flight. As would be done in future flight experiments, that honor would go to animals.

The Soviet Union and the United States both used animals on space flights before sending humans. Dozens of monkeys, dogs, mice, and other assorted animals were used to test initial flights in their space programs. In 1947, fruit flies were the first creatures to enter outer space, going up 68 miles. They survived the trip. However, many other animals gave their lives to the space program. Stress, over-heating, or simple rocket failure resulting in a crash took many of the animals. In 1957, the USSR sent a Russian dog named Laika up in Sputnik II. Laika became the first animal to go into orbit. Ham, also known as "Ham the Astrochimp," was sent up by NASA just three months before Alan Shepard became the first American in space. The Montgolfier brothers did much the same by sending three animals into the sky before allowing a human to go up.

A wicker cage was attached to the bottom of the balloon. Doubtless not knowing what they were getting themselves into, a sheep, duck, and rooster were loaded into the cage and would be the first animals to experience balloon flight.

After the fire was lit, the balloon filled in minutes. Three cannon shots boomed as it rose from the ground. Things got off to a rough start. Just after its release, with more of the recent bad luck with weather, a wind gust tipped the balloon over. It temporarily lay on its side, but quickly righted itself and rose upward. According to two astronomers who were taking measurements and making calculations, it rose at least 1,500 feet in the air. The flight lasted eight minutes, but not as long as the twenty-minute flight Étienne was hoping for.[23] He theorized that was because of the loss of hot air when it tipped over at the start of the flight. Several chased after the balloon and found it lying on the ground with the envelope partially hung up in a tree. However, a Pilâtre de Rozier had already gotten there ahead of them. This would not be his only first. In a few short weeks, he would be aloft himself as the first aeronaut, as these early balloonists would soon be called.

The animals survived the trip almost unscathed. By the time the pursuers of the balloon got to them, they were out of their cage, which had broken on impact. The duck was somewhat injured, but several witnesses said that it had been kicked by the sheep before the flight had even begun.[24]

Meanwhile, at the launch site, there had to have been great relief and excitement. Étienne met with the king and queen inside the palace afterward, and that evening had dinner with members of the Academy. Étienne's wife, Adélaide, was back in Annonay, having just given birth to their daughter the week before. That evening he wrote to her, describing the day's flight. "At one o'clock, we set off a round of ammunition and lighted the fire. Two or three puffs of wind raised doubts about the feasibility of the experiment. However, by dint of muscle and gas, we overcame all obstacles. The machine filled in seven minutes. It was held in place only by ropes and the combined efforts of fifteen or sixteen men. A second round went off. We redoubled the gas, and at the third round, which I may have had fired too soon for fear that the wind might come up and interfere, everyone let go at once. The machine rose majestically, drawing after it a cage containing a sheep, a rooster, and a duck. A few moments after takeoff, a sudden gust of wind tilted it over on its side. Since there was insufficient ballast to keep it vertical, the top afforded the wind a much larger surface than the part where the animals were. At that instant I was afraid it was done for. It got away with losing about a fifth of its gas, however, and continued on it way as majestically as ever for a distance of 1,800 fathoms where the wind tipped it over again so that it steeled gently down to earth."[25]

Initial flights by the Montgolfiers were viewed by only a handful. The flight in front of the Versailles Palace was viewed by thousands, including the king and queen. A structure below the balloon contained the fire that heated the air, sending aloft the balloon, along with a duck, sheep, and rooster.

Man Flies

Jean-François Pilâtre de Rozier, known as "de Rozier," had already made it abundantly well known that he desired to be the first to fly. He seemed like a good candidate based on his background. Born in Metz, France, some two hundred miles east of Paris, he came to Paris at the age of 18 to further his studies. He had an interest in science and had even established his own museum of science. Possessing qualities not usually connected in a single individual, de Rozier had the distinction of being part scientist, part showman, and part daredevil. He made scientific demonstrations before crowds at his museum, inhaling hydrogen, then lighting it on fire when he exhaled.[26] He had developed a respirator which was a forerunner of later gas masks.[27] Using something that looked like a snorkel attached to an air tank, he performed tests in which he would purposely breathe in foul smelling gasses, lying in a trench of excrement for up to half an hour so he could test the tolerance of his lungs while using his device.

He had accomplished a great deal in his 26 years, but now his overwhelming desire was to be the first to fly. He had connections within the royal family which certainly helped him in his quest. King Louis, however, not wanting to risk an innocent life, was a proponent of sending up one of the city's prisoners, with the caveat that a successful flight would result in a pardon. De Rozier was determined, however, and, perhaps only because of his persistence, was granted the honor.

The balloon was again manufactured at the Réveillon factory and, in fact, would be named *The Réveillon*. Its design was more extravagant than any of its predecessors. The background was a bright blue with gold lions, eagles, fleur-de-lis, suns, and signs of the zodiac, along with the king's monogram. It was the largest balloon built to date with a height of 74 feet.[28] It was dazzling.

Flights in the past had been limited by the fact that the source of the lift was earthbound. This was true whether hydrogen gas or heated air was being used. This time, however, the source would be taken with them as they flew. A metal brazier that could be loaded with any burnable substance would hang from the bottom of the balloon and accompany the balloon as it rose. There would now be some control of the altitude, and the life of their flight could be extended for as long as their fuel lasted.

There was a donut-shaped promenade encircling the base of the balloon. It was three feet wide so a person could walk the full circumference, although doing so would seriously upset the balloon's balance. If the flight had two passengers, they would stand on opposite sides of the promenade. If there was only one passenger, a weight could be placed on the opposite side. In addition to people, the promenade would hold pitchforks, straw, buckets of water, and sponges attached to sticks. The empty space in the middle was for the brazier which hung from the neck of the balloon. If the two passengers were standing, they would not be able to see each other due to the bottom of the balloon being between them. There were two holes placed on the inside wall of their circular walkway, 180 degrees apart, so the balloonists could crouch down and look across and at least have some sense of how the other was doing.

There were multiple test flights on the first day. All were tethered. De Rozier went up but was quickly brought down. This is when the imbalance issue was discovered. A 110-pound weight was put on the opposite side from de Rozier, and the test resumed. The balloon then successfully climbed 84 feet and stayed there almost five minutes. Just as the balloon was landing, de Rozier impetuously jumped out of the basket before the balloon was secured. This sent it quickly back up, but the handlers were able to haul it back down.

There were other flights that day. De Rozier was able to get used to controlling the balloon's height by building the fire up or allowing it to die down. There was another test two days later. There was only one flight that day as extremely windy conditions caused the balloon to lean at a 45-degree angle when the tethers were fully extended. In his letters, there is some evidence that Étienne also went up in at least one of these flights, despite the fact that his father had given his blessing to their experiments as long as they never put themselves at risk by going up in a balloon. If he did fly, it seems this was the only occurrence.

The crowds grew each day, even though the flights were not advertised. The next set of flights had 2,000 onlookers. The spectators saw four flights that day, each one higher than the one before. De Rozier soloed the first two. For the third flight, he was joined by a Girond de Villette on a flight that lasted nine minutes.[29] On the fourth flight, de Villette gave way to the Marquis d'Arlandes an infantry major.

Étienne later described the flight in a letter he wrote to Joseph. "...I had two people get in the machine which was held by four ropes 300 feet long. The machine rose their full length and held steady for a quarter of an hour. The people in the gallery let it descend and raised it again several times in succession, and even succeeded twice in adroitly manipulating it so that it nearly grazed the ground and then rose again without touching down at a moment when the wind had carried them out of plumb with the garden. They came down in a neighboring garden above a tree which brushed against the machine, but that accident only served to show that exposure to such risk was not very dangerous and that the machine could be lifted without returning to earth. Thus, we have perfected the machine as far as we could have hoped with the means at our disposal."[30]

The Montgolfiers wanted to be the first to achieve manned, untethered flight. They were in a bit of a hurry because they knew the Robert brothers were preparing another gas balloon and could possibly put a man in the skies before them. But they felt they were ready. October 20 would end up having too much wind and rain to fly. The next day, the balloon was transported from the Réveillon factory to a royal palace, Chateau de la Muette, which was, at that time, on the western outskirts of Paris. This day was another windy one. During a tethered test flight, the wind ripped the balloon open in several places and it fell to earth. As it fell, the contents of the brazier caught the balloon on fire. Upon landing, members of the crowd helped put out the fire before it could completely consume the envelope. Several female seamstresses came to the rescue, and in two-hours' time, repaired the tears in the fabric.

There was a period of calm which seemed right for a launch. They took their places—de Rozier on one side and d'Arlandes opposite him. All was ready. At 1:54 in the afternoon, the ropes holding the balloon were cut and it rose into the sky.[31] After millennia of dreaming of it, man could fly, not attached to the earth in any way. D'Arlandes later commented that he was surprised by the complete lack of noise as they rose.[32] The crowds were soon out of hearing, and although there was a wind that day, there was none for the two balloonists. Balloons are carried to wherever the wind is taking them, so the aeronauts were riding in a sea of calmness. After a short time, they started drifting back down to earth. D'Arlandes was so mesmerized that he forgot to tend to business and de Rozier had to yell at him to feed the fire. Using a pitchfork, D'Arlandes added straw to the fire and their ascent was renewed.

Their height could be adjusted because a brazier was attached to the balloon, but this obviously could cause serious problems as well. They were all alone with fire directly below and a flammable fabric directly above. Because of this, de Rozier was not only the first to fly by balloon but also would one day be its first fatality. He would survive this flight, however, due to some quick action on the part of the aeronauts.

Sparks from the brazier did land on the envelope and started to burn through. The sponges were used to deal with those damaged spots, but a worrisome amount of harm had already been done. After a flight of 25 minutes, de Rozier decided to put down. They successfully landed among windmills, five miles from their starting point.[33] They still had plenty of straw left on board and undoubtedly could have gone farther, but considering the balloon's condition, they thought better of it.

D'Arlandes later described the last few minutes of the flight. "But the intrepid Rozier, who never lost his head, and who judged more surely than I, prevented me from attempting to descend. I then threw a bundle of straw on the fire. We rose again, and another current bore us to the left. We were now close to the ground, between two mills. As soon as we came near the earth, I raised myself over the gallery, and leaning there with my two hands, I felt the balloon pressing softly against my head. I pushed it back and leaped to the ground. Looking round and expecting to see the balloon still distended, I was astonished to find it quite empty and flattened. On looking for Rozier I saw him in his shirt-sleeves creeping out from under the mass of canvas that had fallen over him."[34]

De Rozier would fly again. This was, however, the only recorded flight for D'Arlandes. He would serve in the French army during the revolution, although he would be dismissed on an accusation of cowardice.[35]

The Montgolfiers were right to be concerned about the possibility a gas balloon could win in the race to put a person into the air. Only ten days after their flight, a manned hydrogen balloon was ready. This balloon had several innovations, and many of them would be permanent improvements. The top of the balloon had a valve which could be opened to let out gas. In the future, if a balloonist felt he was ascending too quickly, or wanted to descend, he would be able to manually let gas out. This was done by use of a cord that went from the valve, through the middle of the balloon, to the basket, where it could be pulled by the pilot.

The previous hydrogen balloon was an enclosed envelope. The base of the balloon would now be an open neck.[36] This would allow the balloon to self-regulate, to a certain degree, the amount of gas it contained. Heat from the morning sun or the sun coming from behind a cloud, could heat the gas in the balloon, causing it to swell. The balloon could also swell due to a decrease in the outside air pressure as the balloon gained altitude. Some of the hydrogen gas within the balloon could simply escape through the opening at the bottom of the balloon.

A net now covered the top half of the balloon and was connected to a ring that encircled the balloon at its equator. The passenger-carrying basket would be attached to this ring rather than to the balloon itself. This new feature also gave more structural integrity to the balloon.

In the near future, there would be additional innovations. One of these was a trail rope. A trail rope was simply a heavy rope with one end in the basket and varying amounts of the other end lying on the ground. This acted as reusable ballast to adjust the elevation of the balloon. For example, if the balloon started to descend unexpectedly, the aeronaut could let out a greater amount of rope. With more of the rope resting on the ground, the entire balloon structure would weigh less.

A rip panel was an innovation in which a large section of the balloon could be manually torn away to cause a rapid descent in the case of an emergency. It could also be used to quickly deflate after landing if a strong wind were to come up and potentially drag the balloon across the ground.

The basket for the second manned flight was shaped like a boat, and in fact, was referred to as a gondola. (It would still be a while until the basket would have its recognizable square box shape.) The balloon itself was 26 feet in diameter with vertical red and yellow stripes.[37] It was not as large or as showy as the balloon that took Pilâtre into the air just the week before, but was impressive nonetheless.

The Robert brothers had made advances in how to load the envelope's hydrogen. It was still a process of running sulfuric acid over iron filings. However, this time there would be several stations feeding a central collecting tank and then piped into the balloon. This method, in addition to using a lower concentration of acid, kept the balloon from overheating as happened previously. Their balloon was successfully filled by this method, then moved, while inflated, to its launch site in front of the Tuileries Palace.

There have been occasions in history in which wanting credit for a discovery has led to anger and accusations. That was not the case here. Jacques Charles planned to send a five-foot green test balloon aloft just before the flight. Étienne Montgolfier was a spectator in the crowd that day. Charles walked over and, handing the test balloon to Étienne, declared, "It is for you to show us the way to the skies."[38]

Jacques Charles and Nicolas-Louis Robert found their places at opposite ends of the eight-foot gondola. A barometer, thermometer, telescopes, and sand for ballast went along as well. At 1:30 p.m., the sound of cannon announced their departure. They, along with their innovations, were able to stay aloft for two hours, traveling 27 miles and landing near the town of Nesle.[39] After a soft landing, Charles had Robert get out of the gondola. It is unclear if this request was preplanned or an impulsive act in the moment. Now lighter, Charles and his balloon shot upward quickly. He was the first person to solo and also claimed to be the first to experience two sunsets in a single day.

Their flight together, according to their barometer, had reached approximately 2,000 feet in altitude.[40] For his solo flight, Charles took only ten minutes to soared to a height of 10,000 feet.[41] He stated that he went from spring to winter. His rapid ascent was rough on his body, though. He had "an extraordinary stab of pain" in his right ear and jaw.[42] At that point, he released the top valve and made another soft landing. Whether it was from the acute pain he felt, or that he had simply had enough, this was Jacques Charles' only flight.

This was the last flight of 1783. Incredibly, flight had only begun the previous June, but in this short time, the world had become captivated. Practically anything that could, bore the image of a balloon. People flocked to see each launch. The Charles balloon's second flight drew an estimated half-million people. This sensation had a name—"Balloonmania." In a letter to an acquaintance, Benjamin Franklin wrote, "All Paris was out"[43] and "We think of nothing here at present but of flying. The balloons engross all attention."[44]

The reality of flight spawned books and plays. Since dew seemed to magically evaporate into the air, the play *Cyrano de Bergerac* told of capturing dew which then propelled his balloon skyward as it evaporated. In *The Unparalleled Adventure of One Hans Pfaall*, Edgar Allan Poe wrote a series of newspaper articles that came to be known as *The Balloon Hoax* in which he describes a flight across the Atlantic Ocean. Charles Dickens described an ascent in one of his *Sketches by Boz* newspaper articles. Jules Verne wrote *Five Weeks in a Balloon* telling the story of a balloon expedition across Africa. H.G. Wells wrote *The War in the Air*. Mark Twain placed Tom and Huck in an airship in *Tom Sawyer Abroad*.

Other countries saw air travel firsthand when French aeronauts made flights outside France. They became stars, and this typically led to that nation's countrymen later launching their own balloons. Perhaps simply due to the centuries-long animosity between France and England, England was one location where ballooning did not initially catch on. After King George III expressed interest in the new endeavor, he was informed by England's Royal Society that "no good whatever"[45] could come of it. A newspaper advised Englanders to, "laugh this new folly out of practice as soon as possible."[46] When the first flight from English soil did take place, that pilot was Italian. Perhaps if the pioneers of flight were any other than the French, English aeronautics might have developed earlier. In time, England would have its share of balloonists, but from the earliest flights, and well into the next century, the French would lead the way.

For someone standing on the ground, the sight of a giant balloon drifting overhead would be thrilling enough. But what must it be like to actually fly? Over a hundred years later, Wilbur Wright would be in France to interest French businesses or the French government in his and his brother's flying machine. While there, he was invited to take a flight in a balloon. His account of flight must have been much the same as the original aeronauts. "Once above the treetops, the narrow roads no longer arbitrarily fix the course. The earth is spread out before the eye with a richness of color and beauty of pattern never imagined by those who have gazed at the landscape edgewise only. The rich brown of freshly turned earth, the lighter shades of dry ground, the still lighter browns and yellows of ripening crops, the almost innumerable shades of green produced by grasses and forests, together present a sight whose beauty has been confined to balloonists in the past."[47]

The Giant Balloons

After the initial flights established the balloon's viability, there was a desire to make use of, or at least make money from, these machines. Fees had been charged for people to have an up-close view of launches. This became less profitable, though, as patrons began to realize that after the launch, one could see the flight just about as well by being back a bit and with the benefit of not having to pay. Another money-making possibility would be to allow individuals onto flights and to charge them for the privilege. This led to a quest to build larger balloons that could carry passengers—the more the better.

The first attempt at giant balloons was in January of 1784. The Montgolfiers and de Rozier, just months after inaugurating human flight, teamed up to build a massive balloon roughly 130 feet tall and 100 feet across.[48] It would again have a circular gallery with a gap in the middle that held a brazier five feet across. With the large balloon and a large gallery, it could potentially lift several occupants.

It would be constructed in Lyon. The head of government there was Jacques de Flesselles, and because of his support of the project, the balloon would be named *Le Flesselles*. Monsieur Flesselles would later be a government official in Paris. Only hours after the storming of the Bastille, the mob accused him of being a royalist, then shot and killed him. The rioters took his head and the head of the Bastille's commander, set them on pikes, and marched with them through the city.

The construction of *Le Flesselles* would prove difficult. It was the middle of the winter, and the weather did not cooperate. The fabric was heated to counteract the cold, but that caused it to catch fire, resulting in a postponement until repairs could be made. After those issues were taken care of, test flights had to be postponed due to storms.

The balloon was finally ready for flight on January 19. Considering the problems it had, de Rozier felt the planned-for six-person flight might be better with, at most, himself and a couple of others. Four nobles that had expected to be on the crew took exception to this, drew their swords, and let it be known they were to be on that flight. Their perspective was accepted, and they were allowed on. Just as the balloon was freed, one of the young rope holders leapt aboard, adding even more weight. With its additional cargo, the balloon rose very slowly and was aloft for only 18 minutes. In its already weakened condition, the balloon developed a four-foot tear and came down more quickly than it went up. Despite a hard landing, there were no serious injuries. Although it was a shorter flight than planned, the crowd of 100,000 in attendance was enthralled and treated all the flyers as heroes.

This would be the last flight for the Montgolfier brothers. They would continue to research and write on aeronautics, but their active involvement was over. Their primary focus would return to the family paper business. Future discoveries of theirs, such as Joseph's development of a hydraulic pump, were useful in their factories. In 1790, Étienne was appointed by the new National Constituent Assembly as an administrator of the Ardèche region of France. Both continued to contribute to society, while also managing to keep out of harm's way during the revolution. Their brother, Alexandre, was not as lucky. He was denounced by the revolutionaries and imprisoned in 1794. He became quite ill in prison, and though released, died shortly thereafter.[49]

The difficulties of Le Flesselles put a damper on the idea of larger and larger balloons for many years. The next major attempt at creating a large balloon was in 1863. Le Géant, as the name implies, was a massive balloon, the envelope itself being 196 feet tall.[50] It needed to be big because of the size of what it would be carrying. The car was an enclosed box that was 13 by 8 feet at its base and 10 feet tall.[51] It contained six rooms and could carry up to twenty passengers. On top of this structure was a fenced deck for relaxing and enjoying the fresh air. It was commissioned by perhaps the most famous photographer of his day. As celebrities sometimes do, he went by a single name—Nadar. In previous years, Nadar had ascended in other balloons and taken the very first aerial photographs.

The second flight of Le Géant was to be a long one, traveling 400 miles to the east.[52] It was a smooth flight—initially. Over Germany, heat from the morning sun caused the bag to swell, so some of its hydrogen was released, but not enough. Fearing a rupture, additional hydrogen was released. However, this time too much was released, causing the balloon to make a rapid descent. Landing, it continued to move rapidly along, sometimes bouncing fifty feet into the air. Since this was the nineteenth century, there were no airports, automobiles, or telephone lines in the way. However, there were trains, and an approaching one was barely able to get itself stopped before a terrible collision. The balloon itself finally came to rest, but not before almost all the passengers had jumped or were thrown out. There were plenty of injuries, but amazingly, no one died.[53]

A much more pleasant experience was had at the Paris World's Fair of 1878. Passengers rode in a balloon that was perhaps the largest that was ever made. It had a volume of nearly a million cubic feet and was able to carry 52 passengers at a time. Over the course of the fair, at least 35,000 people were able to experience flight.[54]

The designer of that balloon was the engineer Henri Giffard. He previously invented the injector, an important device that could inject a fluid such as fuel into a fluid of higher density—important in the working of internal combustion engines. Giffard was also the inventor of the dirigible, first flown in 1852. *Dirige* is French for "to direct" and its ability to be directed in its flight is what set it apart from the balloons. His dirigible was based on the work of Jean Meusnier almost seventy years prior. Meusnier was a respected mathematician, contributing Meusnier's Theorem to differential geometry. He also held a degree in engineering and worked and cowrote papers with Antoine Lavoisier. Along with his career in science, he spent twenty years in the military. In 1793, General Meusnier was killed at the age of 39 while fighting for France during the siege of Mainz.[55]

Meusnier was present at the first balloon flight of Jacques Charles and the Robert brothers, in 1783. He had gathered volunteers to various locations surrounding the launch site, armed with quadrants and stop watches to collect data regarding the flight. Only months after the very first balloon flights, Meusnier laid the theoretical foundation for what Girard would use decades later for the first dirigible flights. He envisioned a balloon in an oval shape with forward movement supplied by propellers. Inside of a hydrogen-filled envelope would be one or two large bladders, or ballonets. They would be partially filled with regular atmospheric air. By releasing air from a ballonet into the atmosphere, the overall balloon would have a higher percentage of hydrogen, causing it to rise. Alternately, filling a ballonet with outside air would have the opposite effect, causing a descent. This process could be carried out repeatedly, allowing the airship to stay in flight indefinitely without having to release its finite reserve of hydrogen or ballast. In his original sketches, his balloons appear very much in the shape that later dirigibles would take. Ballonets are a feature of today's airships and give them their stability and ability to stay aloft for weeks at a time. Meusnier's concept of keeping dirigibles aloft is the same as is used by submarines. Not called ballonets, but ballast tanks, they are filled or emptied of surrounding sea water to allow them to submerge or rise in the ocean.

Crossing the Channel

From 1066, when William the Conqueror sailed from Normandy to attack England, to 1944 with the Allies going the opposite direction on D-Day, the English Channel has been perhaps the most important 20 miles in world history. It has also been crossed simply because it was there. Approximately 2,000 swimmers have crossed the English Channel. Within a couple of years of the advent of flight, there was a desire to cross the channel by air.

Jean-Pierre Blanchard was born to a poor family in the Normandy region of France in 1753. He was thrilled by the initial flights, which led to him making several successful flights of his own. Wanting to make a name for himself, he decided that being the first to fly across the English Channel would be an excellent pathway to fame. Blanchard was a small man, making him, physically, well-suited for flying. He was also a driven individual, seemingly another plus, although many would say he was driven to the point of being egotistical and difficult to work with.

John Jeffries, his partner for the cross-channel flight, experienced this difficulty firsthand. Jeffries, a physician, was originally from Boston, and there attended Harvard University. At the outbreak of the American Revolution, he treated injured soldiers, including those injured at the Battle of Bunker Hill.[56] He was a royalist and being so in the city of Boston potentially placed him in difficult circumstances. He decided to move to England to continue his education and, doubtless, to be with people with similar political leanings as his. Jeffries agreed to supply the funding for Blanchard's proposed flight as long as he got to go along. Jeffries liked the idea of being a part of flight, but he also had a great interest in science, especially in the area of meteorology. As it turned out, he would not have time to do much research on the flight.

The mathematics for these flights were worked out ahead of time, so the amount of weight a particular balloon was capable of lifting was known. Fitting with Blanchard's personality, he had a tailor surreptitiously sew lead weights into his clothing.[57] When being weighed pre-flight, he could claim that the total weight was such that his partner would have to stay behind, thus reserving all the glory for himself alone. This ruse was discovered, however, so what now had to be an uncomfortable partnership continued.

Blanchard and Jeffries took off atop the cliffs of Dover during the afternoon of January 7, 1785. The flight went well until they began to lose altitude, most likely from undetected gas leaks. The ballast they carried was thrown overboard, which helped, but was not enough. Everything that could be discarded, was—equipment, scientific instruments, and even coats and pants. It was just enough for a successful landing in France. They settled in some trees a couple hours after they took off "almost as naked as the trees."[58]

In addition to the obvious historical importance of the flight itself, the aeronauts had with them a packet of letters to be delivered in France, making this the first air mail delivery. It was then off to Paris to be honored by the king. Along with praise, Blanchard received an award of 12,000 livres and a lifetime pension. Perhaps because of his British ties, Jeffries received nothing.[59]

Meanwhile, de Rozier, the very first man to fly, was also hoping to be the first to fly across the channel. With the flight of Blanchard and Jeffries, that goal was gone, but he wanted to attempt the crossing anyway. It would not be quite the same trip, though. His plan was to take the opposite path, beginning in France and landing in England. This flight would also be different for another reason. The balloon itself would be a new type. So far, there had been basically two kinds of balloons. The one that used heated air had come to be known as a Montgolfière. The type using hydrogen would be known as a Charlière. The balloons were named, naturally, after those individuals most associated with them—the Montgolfier brothers and Jacques Charles. This new hybrid would henceforth be called a Rozière.

This first Rozière was made up of the two previous types of balloons—a spherical hydrogen balloon on top with a cylindrical hot air balloon underneath. The theory was that the hydrogen balloon would supply the greater lifting force. The hot air balloon below could have its fire tended so that it could make adjustments in altitude without having to release its ballast or hydrogen unless absolutely necessary.

This new endeavor of flight had always been a risky proposition, but with this attempt, even more so. It would be the first time for a flight to carry both open flames as well as the flammable gas hydrogen. Even though the fire and the hydrogen would be separated by the Montgolfière balloon, there was still plenty to be concerned about. De Rozier's fiancé, Susan Dyer, asked him not to go. Even de Rozier himself had expressed doubts about its safety.[60]

The wind currents were generally more favorable in the England to France direction, and de Rozier's flight had to be put off until satisfactory conditions developed. In June they did. The evening of June 15, 1785, a crowd had gathered at Pas-de-Calais to see off de Rozier and a companion, Pierre Romain. The duo departed and drifted toward England. For a short time, things looked good, but the balloon stalled, and then drifted back toward the French shore. And then for reasons not exactly clear, flames came out of the top of the hydrogen balloon. The machine fell to earth some 300 feet from the shore. Both passengers were killed in the crash. They were the first deaths in aviation history. Susan Dyer, de Rozier's fiancé, who was in the crowd, collapsed from the sight of it all and soon died.[61]

Remarkably, the three basic types of balloons were all developed within the first year and a half of ballooning history, and all three types are still with us today. After the channel disaster, the Montgolfière balloon faded from use. Even if the passengers on a trip were unharmed, the Montgolfière, with its attached fire, was usually burned beyond repair upon landing. Not just for safety reasons, hydrogen's further advantage was its greater lifting force; thus, the balloons could be smaller. Because there was no attached fire, the Charlières could usually be reused several times. Hydrogen's big drawbacks, the long wait time and expense of producing hydrogen, soon were not a factor when coal gas became a feature in cities. Before the use of natural gas in the twentieth century, coal gas was piped through cities and used for heating and lighting. It was not quite as buoyant as hydrogen but was cheaper and could be quickly piped directly into a waiting balloon.

Though hydrogen is flammable, with no flame present, disaster is unlikely, although certainly not out of the question. The Charlière would remain the predominate balloon into the twentieth century, until the Hindenburg disaster put a sudden end to its popularity. Gas balloons now only exist in the form of blimps using the heavier, but not flammable, helium. Because helium is not as light as hydrogen, blimps need much more of it, accounting for their immense size. They are still popular today because Meusnier's ballonets make the balloons stable and able to stay aloft for long periods of time, making them popular for advertising and filming stationary objects below, such as football stadiums.

The very first type, the Montgolfière, would make its return. In fact, almost all balloons today are of that type. The brothers' fuel of old shoes and straw has been replaced with propane tanks that sit in the basket and are used to generate the fire that heats the air in the balloon. Despite its horrendous beginning, the Rozière would find its niche. Flights across the Atlantic and Pacific Oceans would be accomplished in Rozière balloons. The advantages de Rozier saw in them in 1785 proved to be what was needed for those long-distance flights, culminating in the first around the world balloon flight in 1999.

Show Men and Show Women

After the initial flights established ballooning's viability, the rush was on to make money. The giant balloons were employed to take on passengers, charging them for the privilege. Balloons would also serve as entertainment, as war machines, and in scientific research.

While there is something dazzling about seeing an enormous balloon take to the air, ways were soon found to make flights even more spectacular. To keep the crowds coming out, a greater visual display or a more dangerous stunt would always have to be in the works. Nothing seemed to be too risky. Thus, flying at night. Then flying at night while shooting off fireworks. Riding in a basket with an animal. Then suspending a platform from the balloon and riding atop a pony, or better, a tiger. Releasing doves during the flight. Ascend in a balloon, then jump out, descending via parachute. Send animals down in parachutes. Descend in a parachute while shooting off fireworks. Suspend a trapeze bar and perform acrobatics. Madame Élisabeth Thible, a well-known singer, performed operatic selections as she floated above the crowd.[62] All those stunts were used in a quest to dazzle crowds.

The parachute would be a key part of the dazzle. It also came to be an important piece of safety equipment and was employed by the early balloonists, out of necessity, many times. Initially, however, it was primarily a gimmick used for its entertainment value. The parachute was an invention that came just months after the first balloon flights. Leonardo da Vinci designed, but did not build, a parachute in the 1500s. His was to be cloth stretched over a frame in the shape of a square-based pyramid. Early attempts all involved some type of solid framework. The Montgolfier brothers experimented with the concept as well. The generally acknowledged inventor was Louis-Sebastian Normand. His first designs were in 1783, the same year as the first flights. He began his experiments by making jumps from trees and tops of buildings. His first jump was with a pair of modified umbrellas, but then he graduated to a single piece of material. He coined the name "parachute" from the French *para* meaning against and *chute* meaning fall.

Jean Blanchard, who had been first to cross the English Channel, knew how to entertain. For example, in 1785, while flying over the streets of Paris, he once sent a dog down by way of parachute.[63] Blanchard found that a solid frame was unnecessary and simply used linen fabric. Without the frame, parachutes now had the advantage of being easily folded and stored in a balloon's basket.

In 1797, barnstormer André Garnerin became the first person to make a high altitude jump from a balloon.[64] While he did make a successful landing, he oscillated greatly as he floated down. When parachutes had been stretched over a frame, the fabric was forced to stay in place. Without that frame, there were stability problems. The French astronomer Jérôme Lalande suggested that this extraneous movement was caused by air trying to escape from underneath the chute—randomly from one side, then another. He suggested a hole be made in the top of the parachute. He thought this would give some of the trapped air a place to go without it causing the oscillations and possibly having the chute simply fold up, sending its passenger rapidly to the ground. Lalande was correct and this did indeed slow the oscillations.

Jean-Pierre Blanchard, in addition to being a part of the first channel crossing, was also the first of these barnstorming showmen. He was the first balloonist in several European countries—Germany, Holland,

Belgium, Switzerland, Poland, and Austria.[65] He was also the first to fly in the new country of the United States. Philadelphia was the seat of government at the time. He advertised his flight for January 8, 1793, with prime seating going for five dollars. Among those who saw the flight that day were President Washington and several people involved in his administration, who incidentally would, in turn, become presidents—Adams, Jefferson, Madison, and Monroe.[66] Blanchard was in America because things were getting uncomfortable in Europe. The French Revolution was well underway, and the rest of Europe had responded, forming a coalition against France. At one point, Blanchard had been arrested in Austria and charged with spreading pro-revolutionary propaganda but was able to escape.

His end came in the middle of a performance. In 1808, while putting on a show in The Hague, he collapsed in his basket after suffering a heart attack. He was brought back to Paris, but his health was greatly impaired, and he passed away the following year.

His wife carried on, perhaps achieving even greater fame than her husband. Sophie Blanchard had accompanied her husband on flights for many years. After his death, she flew alone in spectacular fashion, performing for another ten years. Sophie's trademark look was wearing a long white dress as she flew. Rather than flying in a wicker basket, she stood on a tiny silver gondola which was only three feet long. She specialized in setting off fireworks while flying through the night. She once flew so high that ice formed on her, and she began to bleed from her nose.[67] On another flight, she was trapped in the middle of a hailstorm. She was fearless in the sky. Her exploits in the air are especially surprising because she seemed to be the opposite on the ground. Small and timid, she was easily frightened. Loud noises startled her. She flew through the air on a tiny plank, yet she was afraid of taking rides in carriages. She was unassuming while on the ground, but when up in the air she was daring and a celebrity.

Like Jean-Pierre, Sophie's end would come from ballooning. On July 6, 1819, she rose from the Jardin de Tivoli in Paris. Fireworks close to a hydrogen balloon would seem to be a bad idea, but it was a common part of her act. This time it caught up to her. While performing her usual fireworks display, the balloon caught fire. The crowd below, unaware what exactly was happening, cheered the pyrotechnics. The burning balloon fell and landed on a Paris rooftop. Sophie hung there for a time, entangled in the ropes, but then fell to her death on the street below.[68] She was already gone by the time spectators got to her.

The Garnerin family was another that made balloon entertainment its career. The family consisted of father André, of parachuting fame; wife Jeanne, who was the first female to pilot a balloon alone; and their young niece, Elisa. André, who went by the name Citizen Garnerin during the period of the French Revolution, joined the French army combatting the forces of the First Coalition. He was captured by the British and spent three years in an Austrian prison. During that time, he worked on his design for a parachute.

Napoléon Bonaparte was not someone to annoy, but André Garnerin did just that. Momentous occasions at that time were often celebrated with a balloon launch as part of the festivities. In 1804, to commemorate his ascendency to emperor, Napoléon commissioned Garnerin to launch a decorated balloon with a crown riding on top. In a coincidence difficult to imagine, after rising from the square in front of Notre Dame, the balloon drifted over the Alps to the opposing capital of Rome. There the crown fell off and, improbably, landed on the tomb of the emperor Nero. Bonaparte was not happy with the subsequent Napoléon/Nero comparisons.[69]

Garnerin's wife, Jeanne, was first simply an admirer of André, then his student, then his flying partner, then his wife. She was not only the first woman to pilot a balloon alone but was also the first to descend by parachute.[70] She gave up ballooning after the death of André, passing the mantle to niece Elisa. Elisa continued the family tradition, touring Europe, ascending almost forty times, and typically, descending by parachute.[71]

Warfare

While for many, the primary use of balloons was to entertain, doing so led to advances in flight. Early on, plans were developed for how balloons could be used in warfare. The French thought they might be used to finally defeat the British who had been under a four-year siege at Gibraltar. There were fantasies about their use in an invasion of England.

To assist the war effort, the government established an aerostatic school at Mendon, France. French chemists Nicolas-Jacques Conté and Jean-Marie-Joseph Coutelle headed up the effort, creating, in essence, the world's first air force. There was a team of 26 aeronauts trained on four balloons, with more balloons and balloonists to be added as time went on.[72]

These balloons used hydrogen, although there were difficulties in the production of hydrogen even in the best of conditions. This was even more true now, since a wartime production facility could be under threat of attack at any time. The facility might also have to relocate as front lines shifted depending on how a battle progressed. Additionally, because sulfur was needed to make gunpowder, methods of producing hydrogen by using sulfuric acid had to be found.

The first use of balloons during time of war was in the spring of 1794. France was fighting Dutch and Austrian troops of the First Coalition in Maubarge, in the northeast corner of France. Like almost all the other early wartime uses of balloons, its purpose was to provide reconnaissance on opposition troops. While doing its reconnaissance, pilots also found time to drop propaganda leaflets on the enemy. A tethered balloon went up twice a day, spying on enemy troops that were otherwise hidden by hills, trees, and a variety of terrain. The aeronauts not only supplied information on their movements, but also directed artillery fire toward an unseen enemy. Information gained was sent to those below by use of signal flags, sliding a message down the tether rope, or attaching a weight to a written message and simply dropping it. Opposing troops made history as well by supplying the first anti-aircraft fire.

After the fighting at Maubarge, the French advanced across the border into Belgium. The troops and balloons saw action at several other sites including Fleurus, where balloon reconnaissance was likely instrumental in a French victory.

To the south, Napoléon used a balloon during his Italian campaign, though he never seemed to be a strong advocate of them. When he later took troops and a scientific corps to Egypt, he also took a balloon. However, before it was even unloaded from the boat, it went to the bottom of the sea thanks to Admiral Nelson's naval attack.[73] After returning to France and taking power, Napoléon closed the school at Mendon. It would not reopen until 1871. Napoléon may also have been soured on the use of balloons by Garnerin's Roman misadventure that turned a celebration for him into an embarrassment. He could have profited from better reconnaissance, at the very least at the battle of Waterloo. It would have helped him to know about the Prussian troops that suddenly appeared off his left flank, turning the tide of the battle against him.

Although clearly eclipsed by other forms of flight in later years, balloons would continue to play an important role during times of war. During the American Civil War, Thaddeus Lowe would aid the Northern side in several battles. In 1861, he took off from Cincinnati and drifted all the way to South Carolina, a state not fond of visiting Northerners during those times. Additionally, his arrival was the week after Savannah had fired on Fort Sumpter, beginning the Civil War. He was arrested as a spy but was able to talk his way into being released.[74]

It would probably have been better for the Confederates if they had held on to him. He went directly to Washington where, on the mall, he went up in a balloon. He demonstrated how he could communicate by sending telegraph messages directly to the White House. He met with Abraham Lincoln that night and received his blessing to establish a unit of balloonists.[75] He was attached to General McClellan's command and accompanied him on his Peninsula Campaign.

Balloons were not used extensively but played a role in a number of Civil War battles. First, the military leaders would need to be shown their effectiveness. At one point, Northern general George Custer was taken up in a balloon to demonstrate to him its reconnaissance abilities. Custer, although typically thought of as brash and courageous, was not so during this flight. He chose to remain seated at the bottom of the basket throughout. He later wrote, "To me it seemed fragile indeed … the gaps in the wicker work in the sides and the bottom seemed immense and the further we receded from the earth, the larger they seemed to become." Custer became more concerned as his pilot, "…began jumping up and down to prove its strength. My fears were redoubled. I expected to see the bottom of the basket giving way, and one or both of us dashed to the earth."[76]

During the Franco-Prussian War, Paris was under siege. In response, the world's first airlift took place. Over 60 balloons were launched with almost all of them landing in safe territory, allowing many to make their escape.[77]

Balloons continued to be used during warfare even after the Wright brothers revolutionized flight. Observation balloons were employed in the Boer War, the Spanish-American War, and over the trenches of World War I. Balloons tethered with steel cables were flown over London during the Blitz. German pilots either had to risk being downed if their planes struck a steel cable or to avoid them by flying so high the bombers' accuracy was affected. The same was done on the Normandy beaches to protect the troops after the D-Day landings. During World War II, Japan launched bomb-carrying balloons toward the West Coast of the United States. They were largely ineffective, although one killed six members of a family in Bly, Oregon.[78]

Science

Fortunately, balloons have been used for peaceful pursuits as well. Before crossing the English Channel, Jeffries and Blanchard had made a test flight together. Blanchard was the pilot, with Jefferies gathering scientific data. He was armed with thermometers, compasses, barometers, and various other instruments. Although Jeffries planned on doing so, no scientific measurements were made on the flight across the channel because his scientific instruments were among the many items that were thrown overboard to complete the flight.

The first purely scientific endeavor took place in 1803 by Étienne

Robertson, with a music teacher by the name of Lloest assisting him. As they rose, they observed the changes that took place in magnetism and in the boiling point of water. They found that the temperature at which water boils decreased as they went higher. Robertson noticed that he could put his hand into boiling water when at a high altitude without feeling pain.[79]

In 1804, mathematician Jean-Baptiste Biot and chemist Joseph Gay-Lussac went on their own expeditions. They brought scientific instruments and even carried along small animals to see what effect the change in altitude might have on them. Taking samples of air at various heights, they found no change in its composition as they ascended. When at an elevation of over 20,000 feet, they opened a flask that contained a vacuum, allowing the air at that altitude to flow in. Back on solid ground, they opened the flask under water and observed that it filled halfway, showing the air was only half as dense at that higher elevation.[80] Gay-Lussac noted that his pulse and breathing were greatly increased. They attributed this to the change in elevation, though, it could have been influenced at least partially by the stress of the flight. They also examined the effects of electromagnetic radiation in the upper atmosphere.

The French would continue the pursuit of flight throughout the century. Alphonse Pénaud was born in Paris in 1850, the son of a French admiral. Throughout his life, he would not be able to walk without crutches because of a hip disorder. He had an interest in flight and designed models such as helicopters and ornithopters (airplane with wings that flapped). One of his models presaged the modern airplane. He called it the Planophore, and one day in 1871 in front of the Tuileries, and a crowd of onlookers, it flew approximately 130 feet. It contained several innovations and flew stably. He made plans for full size airplanes but was not able to find funding. Unable to deal with his inability to fund his dreams, he tragically committed suicide at the age of 30.

Though he did not achieve the success he sought, his models were copied by many. Entrepreneurs made children's toys from Pénaud's designs. Milton Wright of Ohio bought one for his kids. His children, Orville and Wilbur, said that toy was the spark that began their imagination of flight.[81]

III

A Revolution in Chemistry

"A revolution in physics and chemistry"[1]—at least that is what Antoine Lavoisier thought he might be approaching as he explored the nature of fire. Though at the time controversial, most scientists ultimately agreed that what he found was, indeed, revolutionary. The science of his day believed fire was basically understood. There was a well-established theory of phlogiston (fire as a material substance) that had wide acceptance in the scientific community. Lavoisier debunked that theory and found the true nature of fire, establishing the link between it and oxygen. In doing so, he turned science upside down in much the same way Copernicus had with his age's theory of the solar system.

Lavoisier stated that substances can exist in three states—as a gas, a liquid, or a solid—established the law of the conservation of matter, and found that our atmosphere is made up of nitrogen and oxygen. He helped create a logical procedure for naming substances, fixing a confusing, contradictory system. He showed that water was not an element but made up of hydrogen and oxygen.[2] His work, thus, put an end to the four-element theory which had been in existence since the time of Aristotle. He discovered the nature of combustion, oxidation, and respiration. He created what was possibly the outstanding laboratory of the day, and showed the importance of careful, precise measurements. Lavoisier has been called the Father of Modern Chemistry.[3]

His contributions apart from chemistry were remarkable as well. He wrote hundreds of papers which advanced science and French society. The papers spoke to improving prison conditions, hospitals, street lighting, drinking water, and the French economy. He was a chairman or a member of many important committees, guiding them despite the turmoil of the times. In doing so, he sought to make a better society by his work in the areas of agriculture, sanitation, education, and the nation's finances. Lavoisier headed the committee that established the metric system. He was the president of the French Academy of Sciences. He established a model farm and used it to teach effective agricultural techniques to French farmers.

And despite all the good he did, on May 8, 1794, the French government sent him to the guillotine, then buried him in an unmarked mass grave.

Antoine's Young Life

Antoine Laurent Lavoisier was born into a bourgeois family in the city of Paris on August 26, 1743. His father, Jean-Antoine, was a lawyer who, over the years, increased his income with good investments. His desire was that his son go into the law as he did. His son did, in fact, get his law degree, but would ultimately gravitate toward the sciences.

Antoine's mother, Emilie Punctis, came from a very well-to-do family that had made its fortune in the meat industry. Two years after his birth, Antoine gained a sister, Marie, though she would die from smallpox when she was fifteen. Antoine married, but did not have any children, so the Lavoisier family line ended with him. Just a year after Marie was born, Emilie, who never fully recovered from Marie's birth, died. Jean then moved himself and his two small children into the home where Emilie's mother and sister, Constance, lived. Twenty-year-old Constance would forego marriage to serve as the primary caregiver and teacher for the children. With Constance's commitment and the family wealth, the children were well watched over.

In 1754, at age 11, Antoine was enrolled in Mazarin College, a well-respected, private, secondary school, where he studied primarily science and law. According to an identification document, he grew to five feet four inches tall, which was an average height for that time. He had chestnut hair and brown eyes. Antoine was already showing a quiet, studious nature and excelled in school. At age 19, he enrolled in the Paris School of Law, and in 1764, obtained his degree in the law, though he never practiced. Although he was not destined to become a lawyer, it trained him for the wide-ranging work he would do in French society and government.

It was about this time that he first inherited money. He eventually received several inheritances, primarily from his mother's side of the family. These inherited monies, in addition to his own investments, would make him a rich man. In the future, his wealth allowed him to purchase the very best scientific equipment and gave him the time to devote to his experiments. However, one day his wealth would be just one more mark against him at his trial.

His fascination with science grew. One of Lavoisier's first steps in that direction came in meteorology. One of his professors encouraged him to take regular measurements of each day's weather and record the data. It was a habit he would continue throughout his life. He had many accomplished teachers that led him toward science as his life's pursuit. In his schooling he had acquired a background not only in chemistry, but also in areas such as physics, mathematics, minerology, and geology. These would all be valuable in the many directions his future life would take.

One of his instructors, and a friend of his father, was geologist Jean-Étienne Guettard. He invited Lavoisier to accompany him on a mineralogical survey of France. He and Guettard spent months travelling throughout France, mapping its minerals, soils, and waters. They later published a book containing their findings, including detailed maps.[4]

Lavoisier presented the first of many papers before the French Academy of Sciences, covering a variety of subjects. He would go on to write reports on the aurora borealis, sanitation, and the formation of the earth. His mentor, Guettard, encouraged him to write *The Analysis of Gypsum*. Gypsum was an important ingredient in plaster of paris, so called because of large quantities of gypsum found in the region around Paris. The Academy sponsored a contest in which they called for papers on the topic of improving street lighting. Lavoisier submitted a 70-page paper and was one of four winners.[5] To gain what he considered the necessary knowledge on the subject, he covered the walls of a room in dark cloth to simulate the nighttime. He spent six weeks in that room, researching various devices and fuels to produce maximum lighting.[6]

He desperately wanted to be admitted to the Academy. However, there had to first be a vacancy, then a positive vote of the Academy, and finally, the support of the king. One of the members did oblige, and his passing opened the needed vacant position. Lavoisier was voted in and, in his mid-twenties, joined others that were typically in their fifties or sixties.

At about this same time, he invested some of his inheritance money into a tax collecting agency known as the General Farm. It was expensive to buy into, and he could initially only afford a partial share and used up most of his inheritance doing so. However, once in, an individual could make millions a year. Lavoisier would become one of the richest men in France.

The General Farm's role was to collect tax money on behalf of the king, with Antoine being one of 40 partners that oversaw the operation. These investors were typically well off financially and employed others who did the physical collecting. Lavoisier had certain supervisory duties within the General Farm. He reported to one of the senior partners, Jacques Paulze, who was a Parisian lawyer and recently widowed. Jacques lived with his daughter, Marie-Anne Pierette Paulze, who was 13 at the time. Marie had received her schooling at the Montbrison Convent. It was common for girls, if they did receive an education, to do so at a convent.

However, Marie's formal education came to an end when she returned home to live with her father following the death of his wife. Lavoisier became infatuated with the blue-eyed, brown-haired girl, and though he was twice her age, married Marie the following year.

Marie was intelligent and attractive, and came to be an indispensable part of Antoine's further work. She learned Latin and English, enabling her to translate the work of other scientists, and to write letters to them for her husband. She was tutored in chemistry, allowing her to be part secretary, part lab assistant, and perhaps it is not a stretch to call her a co-researcher for the next 22 years of their marriage. She took notes and wrote up his experiments, typically sitting at a nearby table as her husband conducted the experiments. She included intricate drawings of the various pieces of equipment and how the whole was laid out. She gained her artistic skill by taking lessons from the premier French artist of the day, Jacques-Louis David.[7]

This painting, by Jacques-Louis David, shows Marie and Antoine Lavoisier in a pose depicting their personal and professional relationship.

There was an amazing collection of scientific discoveries during the revolutionary era. However, while there have been many great artists and authors through France's history, there were few during this time. Jacques-Louis David stood out. Many well-known paintings of that period have David as their artist. *The Death of Marat, The Coronation of Napoleon*, and *Napoleon Crossing the Alps* are all his. He also painted the best-known portrait of Antoine and Marie—a huge six by nine–foot painting that now hangs in the Metropolitan Museum of Art in New York.

David was a supporter of the revolution and, as a member of the National Convention, would cast a vote for the execution of the king. He fell in with the most radical elements of the French Revolution, aligning himself with Jean-Paul Marat and Maximilien Robespierre. After Napoléon Bonaparte's coup, David quickly sided with him. He was not beyond using his art to make political points. In *Napoleon Crossing the Alps*, David chose to show Napoléon majestically astride a white stallion, when in fact, he was more likely riding a mule. He remained a strong supporter of Napoléon; however, when invited by him to be one of the savants going on an expedition to Egypt, he chose to stay in France. After Napoléon's final defeat and the reestablishment of the monarchy, David left the country. In 1825, he was struck down by a carriage in the streets of Brussels. He was buried in Brussels, his body not being allowed back into what was by then monarchist France.[8]

Alchemy

Alchemy was the precursor of chemistry. It has its roots 2,000 years ago, while what legitimately might be called the science of chemistry did not really take place until the eighteenth century. Alchemy is related to chemistry as astrology is to astronomy. Many of the "discoveries" and theories of the alchemists consisted of wrong information which would cause centuries worth of confusion. By comparison, other endeavors, such as geology, oceanography, anatomy, and geometry did not generally suffer from this kind of false information clouding its way. Those areas were blank slates, open to development without having to undo false information.

In their laboratories, they boiled, they smelted, and they distilled, seeking to create a valuable metal or medicine. While alchemists posited much incorrect information, as they dabbled, they were setting the stage for future discoveries. There was the fruitless search for a magical philosopher's stone, but there was also the discovery of how to create various acids and medicines. It would be left to scientists such as Lavoisier and Priestley to separate the wheat from the chaff.

Adding to the confusion was that, for various reasons, much of this alchemic knowledge was secretive. Information gathered was written down in codes and symbols. Greed was undoubtedly one of the reasons for this. The quest for gold was a driving force in alchemy, but what would be the value of gold if everyone knew the secret to making it?

Another reason for secrecy was that alchemy was often held in disfavor by religious and political leaders. In 296 CE, Diocletian banned alchemy in the Roman world.[9] He was likely concerned about the effect on the Roman economy if individuals could make gold at will. Alchemy was prohibited by powers such as Pope John XXII in 1317 and Henry IV in 1404 either because it was seen as being a sinister, cultic affair or, again, for economic reasons.[10]

Alchemy seems to have had its beginnings in Egypt, and specifically in the city of Alexandria. In 332 BCE, Egypt was conquered by Alexander the Great who founded the city, giving it a Hellenistic influence along with his name. Its lighthouse was one of the Seven Wonders of the Ancient World, and the Library at Alexandria was the largest in the ancient world. It would become the world's most populous city, and because of Alexander, its influence reached throughout much of the known world. Because it was inhabited by scholars and by craftsmen in the areas of metallurgy, dyeing, and glassmaking, it made Egypt a prime location for alchemic pursuits.

As they explored the makeup of the world, Greek scholars developed a number of theories regarding matter. Some would last and some would not. Empedocles theorized that the world was made up of four essential elements—earth, air, fire, and water.[11] Those four substances, and various combinations of them, were the building blocks of all matter. Aristotle agreed, although to the four earthly elements, he added a mysterious substance he called ether, which was the domain of the planets and the stars. Though disputed by some, the theory that the world was made up of these basic elements would hold sway for 2000 years.

At about the same time, approximately 400 BCE, Greek philosopher Leucippes theorized that all matter was composed of tiny particles. Leucippes' theory, refined by his student Democritus, stated that these particles last forever, but could differ by size, shape, and weight. Democritus called these particles *atamos*, Greek for uncuttable (irreducible).[12] He believed that substances could be formed by similar atoms or combinations of different atoms. It would be many years before this atomic theory gained universal acceptance.

Alchemy's primary aim was to find precious metals. Just as it could be observed that water becomes ice and caterpillars turn into butterflies, alchemists believed that metals, while still in the ground, might transform to become other substances—perhaps silver or gold. They believed that this transformation was something that happened naturally, but the alchemists thought they might be able to artificially speed up the process with the aid of something called a philosopher's stone. It was never found because such a thing never existed, however, that did not stop some from describing what it might look like. Some said it could change colors. When white, it could turn a substance to silver, and when red, to gold. It perhaps glowed. Some conjectured it was not necessarily a stone at all, but maybe a thick powder.

Another aspect of alchemy was a search for substances that could cure disease. There was a quest to find what was called a "panacea" that could cure any ailment. The ultimate goal was to find the so-called "Elixer of Life" that would give a person eternal life. While those substances did not exist, the search for them would lead to the science of pharmacology.

Many of the Greek advances flowed into the Roman Empire. Both empires excelled in literature, architecture, warfare, and medicine. However, Roman mathematics and science made little progress, being crippled by its cumbersome number system. Roman numerals may be adequate for simply counting numbers of objects, but the system is a nightmare for even simple addition, let alone any more advanced operations. In the Roman Empire, and thus most of Europe, science entered a period of decline lasting to the time of the Scientific Revolution, hundreds of years away.

While Europe was dealing with its Dark Ages, the Arabian world was making advances in mathematics, astronomy, and alchemy. They coined the word *khemia*. The exact meaning of the word seems to be lost to history, but combining it with the Arabian word for "the," "al" gave us the name alchemy (other Arabian words beginning with "the" include algorithm, algebra, and alkali). The greatest alchemic advances during this time were in the Arabian world, but there were other important discoveries, such as gunpowder in China and various medicines in India.

Around the twelfth century, the light began to dawn in Europe. Paracelsus, Albertus Magnus, Roger Bacon, and Nicholas Flammel were important alchemists during this time. They, and others, made discoveries that would turn out to be scientifically important. They discovered minerals such as phosphorus and zinc. They improved distillation practices. Acids, such as sulfuric and nitric acid, were found in the fourteenth century. Alchemy led to the development of perfumes, dyes, glass, and medicines.

Though sometimes they bore fruit, many of the alchemic finds were misguided or simply wrong, and much time and energy was wasted. There were brilliant people, including Thomas Aquinas, Francis Bacon, and Isaac Newton, who spent years looking for objects that did not exist. Especially puzzling is the case of Newton who, while discovering secrets of the universe, also wrote over a half-million words on the subject of alchemy.[13] He found the secrets of light and gravity, the laws of motion, and calculus, but was also looking for objects such as the philosopher's stone. While Newton may have been the greatest scientist that ever lived, notes found well after his death showed he spent an enormous amount of time over the course of three decades pursuing non-existent substances and wrong theories.[14]

It turns out that the quest to change one element into another was not an impossibility after all. What makes an atom one of a particular element is determined solely by the number of protons in its nucleus. To change the number of protons is to change the element. During radioactive decay, the nucleus of an element, such as uranium, breaks up and yields other elements.

Platinum atoms contain 78 protons and gold has 79. If there were a way to squeeze one more proton in, atoms of platinum would become atoms of gold. Scientists in fact have been able to obtain small amounts of gold in the laboratory. It is nice to know that the alchemists were not totally misguided in their quest. The science of their day, though, was incapable of changing nuclei. Most importantly, however, alchemy ultimately led to the science of chemistry.

Pioneers of Chemistry

There is no exact point, nor is there a single individual, that led the transition from alchemy to modern chemistry, and the transition was not an even one. Many scientists continued to live both in the world of alchemy and the world of chemistry. Some made discoveries important to the establishment of chemistry while, at the same time, maintaining a search for magic elixirs and other spurious substances.

Many made important discoveries that led to the work of Antoine Lavoisier. Men such as Robert Boyle, George Stahl, Joseph Black, and others made significate breakthroughs. Progress, though, was sometimes difficult. One stumbling block was communication. While some were quite secretive, some of these natural philosophers (the word scientist was not yet in common use) freely shared information regarding their discoveries. However, try as they might, this sharing was often difficult due to distance and language differences. Also, until the end of the eighteenth century, there was a wide variety in how substances were even named.

It is often difficult to assign appropriate credit for discoveries. In addition to the issue of communication, individuals sometimes waited years to publish their results. Also, often more than one chemist was working on a particular problem, with discoveries made at roughly the same time. Citing credit was also difficult because some of these early chemists might indeed make a breakthrough, yet did not really know what they had, or completely misidentified it. Those cases were somewhat like that of Christopher Columbus who "discovered" a route to the Far East but was actually a half world away.

Of earth, air, fire, and water, these natural philosophers' primary interest was in that of air. For centuries, it had attracted the least attention of Empedocles' elements. Air was simply there, never changing, never very noteworthy. However, it would soon become the source for unlocking the secrets of all matter and would lead to the beginning of modern chemistry. No one person, but a series of men, led the way.

Robert Boyle was one of the first to begin this transition from alchemy to chemistry. His was born in Ireland in 1627 and lived most of his life there. He came from a wealthy family, which was a background he would have in common with several of these men. Wealth gave them the ability to purchase equipment, to set up their own laboratories, and allowed them the leisure time that could be devoted to research. When he was eight years old, and following the death of his mother, he was sent to Eton College in England. There he embarked on a literary career, primarily focused on religious topics, but in 1649 began his scientific experiments.

By combining sulfuric acid and iron filings, he created what would become known as hydrogen. Though it was a mystery to him at the time, it was a possible crack in the four-element theory. Aristotle's view was that the air we breathe was a single element. If this new gas was also a naturally occurring substance contained in the common air that surrounds us, air could not be a single element. While it was not yet totally clear to him, this gas, and each new gas found in the future, was another step in the demise of the theory that the world was made up of a very small handful of elements.

In 1661, he wrote a ground-breaking book, *The Sceptical Chymist*. Unlike much of the purposely confusing writings of the alchemists, Boyle's book was meant to be understood. Rather than simply theorizing, it emphasized experimentation to establish and prove hypotheses.

Later, together with fellow Irelander Robert Hooke, he established Boyle's Law stating that the pressure and volume of a gas are inversely proportional. He was also a proponent of the world's being made up of tiny particles. However, he assumed those tiny particles were identical, with different substances formed by rearrangements of the particles. He, unfortunately, also continued the search for the philosopher's stone. Despite some steps in the wrong direction, Robert Boyle has been considered by many to be the first true chemist.

George Stahl was born in 1660 in what is now Germany. He attended the University of Jena to study medicine and stayed in that field throughout his life, both as a teacher and a personal physician. Part of his contribution to science was his experimentation with potential medicines. Like Boyle, he was an atomist—a believer in the concept of matter being made up of small particles. For much of his life he was a believer in alchemy, although over time, he grew to be skeptical of it. He developed the popular, but ultimately wrong, theory of phlogiston. Phlogiston was a mistake that remained a dominant theory for almost a century. It was first advanced by Johann Becher in 1699, and Stahl credited him for its inception, but it was Stahl that developed the theory to its full extent.[15]

The concept of phlogiston explained the nature of fire. Fire was a dramatic example of change. If a pile of wood was lit on fire, the result was flames, smoke, and ultimately a small pile of ash. What exactly was taking place? Furthermore, why did some substances burn, some did not, and some burned only a little? Becher felt flammable objects contained a substance that Stahl would call phlogiston, a word that was taken from the Greek, meaning "set on fire." A substance such as wood must have a lot. Iron seemed to have none. When a piece of wood burned, the phlogiston it held was set free and dissolved into the air, and it burned until all its phlogiston was used up. The remaining ash is simply wood without its phlogiston. That a pile of wood burned down to such a small pile of ash was further verification that something must have escaped. What happened during calcination (a process of heating substances to very high temperatures) also seemed to fit into his phlogiston theory. This theory would last for decades until Antoine Lavoisier proved it to be totally wrong.

Joseph Black was born in Bordeaux, France, in 1728, though his family was from Scotland. He explored the fields of art, then medicine, and then chemistry. He attended the University of Glasgow and became a professor there and later at the University of Edinburgh. Joseph never married, but he was a popular professor, and had friendships with members of the Scottish Enlightenment, such as economist Adam Smith and philosopher David Hume. Black was not as well known as he might have been because he chose not to publish many of his findings. He did much of his research simply with the goal of benefiting Scottish industry.

In 1745, Black discovered carbon dioxide, or as he called it, "fixed air" by pouring acid on limestone and collecting the air that bubbled up.[16] It was called fixed air because he found that it could be absorbed, or fixed, by strong bases. Black found that this air had different characteristics and was convinced it was different from the common air that they all breathed. This air (they would not regularly be referred to as "gasses" until years later) weighed more than common air. Substances that typically were flammable did not burn if they were in a container surrounded with this new air. An animal (mice were usually his test cases) could not breathe when in it. This obviously was not common air.

Henry Cavendish, like Joseph Black, was not French despite being born in France. He was born in Nice in 1731 where his English family was living at the time. They were rich, and he would soon inherit a substantial amount of money. He attended Cambridge University. An extremely shy man, he published no books and very few scientific papers, yet made significant contributions to science. In 1798, Cavendish devised an ingenious experiment in which he closely estimated the density, and then the mass, of the Earth to within 1 percent its actual value.[17] In 1766 he discovered hydrogen. Robert Boyle had found it five years before but did not quite know what he had and did little with it. Cavendish saw it as being a new element and is thus usually credited with its discovery. He had found the fuel that would lift the Charlière balloons in the future. He called it "inflammable air" because of its ability to easily be lit on fire. Unfortunately, too many balloonists would come to find out it was an appropriate name.

Daniel Rutherford was a student at the University of Edinburgh, as was his nephew, the author Walter Scott. Daniel's father and the chemist Joseph Black were both professors at the university while he attended. While there, in 1722, Rutherford was credited with the discovery of nitrogen.[18] Although Henry Cavendish and Joseph Priestley had found it at roughly the same time, Rutherford was the first to publish. He referred to it as "noxious air" or "phlogisticated air." He called it noxious air because a mouse placed in a container with only this type of air to breathe, quickly died. Wood, or other typically flammable objects, could not be made to burn when surrounded by only this kind of air. Rutherford assumed that must be because this new air was already full of phlogiston. With no room for more, there could be no fire. Therefore, the name phlogisticated air.

Carl Schleele was born in Sweden in 1742 and lived there his entire life. He became a pharmacist, which gave him access to chemicals, but he did not have an advanced, expensive laboratory like many others. He wrote only one book and did not keep particularly good notes on his work. Despite this, he isolated a great number of substances—manganese, barium, citric acid, chlorine, and others. His most important find was the gas now known as oxygen.[19] He discovered it at approximately the same time as, but independently of, Joseph Priestley. Schleele found it to be flammable and thus gave it the name "fire air."

Because it was flammable, this new air must not contain any phlogiston. Schleele continued to be a proponent of the theory of phlogiston throughout his life. That, though, is perhaps only because he lived a short life, which ended just as the theory was beginning to fall into disfavor. He was only 43 when he died. The symptoms he showed imply that perhaps a lack of the best of equipment and his work with dangerous chemicals such as mercury, arsenic, and cyanide led to his early death. In his final days, he married the widow of a fellow apothecary so his laboratory would stay with someone that would care for it.

Joseph Priestley

These chemists lived at roughly the same time, and all made important contributions, but Joseph Priestley was the one who had the most direct impact on Antoine Lavoisier's life. Priestley was English and Lavoisier was French. One was poor and the other rich. However, in other ways, their lives mirrored each other. Both lost their mothers while they were children and ended up living with aunts. Both faced harassment from their countries. Priestley ultimately left England, going to America for his own safety. He departed England forever on April 8, 1794, when, at the same time, Lavoisier was in prison and had one month before his death at the hand of the French government. Though they reached different conclusions regarding its true nature, they were the scientists that led the way in unlocking the secret of fire.

Priestley was a controversial and important man. He dined with President Washington. He was friends with Benjamin Franklin. In fact, it was his account that let the world know of Franklin's stormy kite flying adventure.[20] There were many years of correspondence between former presidents John Adams and Thomas Jefferson. Adams' and Jefferson's friendship struggled for a time, but they patched up their relationship in the last years of their lives, exchanging 165 letters. Author Steven Johnson speaks to Priestley's importance by citing that in those letters, Benjamin Franklin is mentioned five times, George Washington three times, Alexander Hamilton twice, and Joseph Priestley 52 times.[21]

Priestley was a minister who had chemistry as a sideline. He did not enter into chemical research until he was in his mid-thirties, yet made major discoveries. He coined the term "rubber" when he saw the substance could rub out writing errors.[22] His typical method of exploration was to pour various acids on various metals and wait to see what bubbled up. He isolated ten different gasses, among them, sulfur dioxide, nitrogen dioxide, nitrous oxide, and ammonia. Once isolated, though invisible, these airs could be distinguished by various tests. Measuring a gas' density or testing to see if it was flammable, breathable, or soluble in various liquids allowed these early chemists to distinguish between various gasses and determine if something truly new had been found.

In 1767, Joseph moved to a new congregation in Leeds where his housing happened to be next to a brewery. Vats of brew would emit carbon dioxide as part of their natural process of fermentation. He found that water, when slightly agitated and held over the vats, absorbed some of that carbon dioxide. Later, he discovered the gas could be directly injected into the water. Thus, carbonated water was born.[23] Priestley thought it might be a cure for scurvy. It was not, but it was the start of the trillion-dollar soft drink industry.

In August 1774, he made his most important discovery, that of oxygen. When heated or calcined, mercury produces what was then called a calx, now called an oxide, on its surface. All metals produce a calx.

Mercury, though it is a metal that is liquid at room temperature, does the same. Priestley placed a sample of mercury in a sealed container, used a lens that was a foot in diameter and the strength of the sun, to heat the mercury until a red calx appeared on its surface. Then, when heating that red calx, it disappeared, the mercury was restored, and a gas was emitted. He assumed the calx was simply mercury that had lost its phlogiston. This gas, however, was something new to him.

He had discovered oxygen, though, initially, he wasn't quite sure what he had. He thought it might be an altered form of common air, as ice is simply water in an altered state, but still the same substance. Unknown to Priestley, Schleele had already discovered this gas, but failed to publish the fact, and by the time he did, credit had been given to Priestley. What Scheele called fire air, Priestley called dephlogistated air. That was because, by Priestley's thinking, an object burns only when it can release its phlogiston. This phlogiston, however, must have somewhere to go. This new air that supported combustion must have the room for it, and that only happens if the air is not already filled with phlogiston.

This gas had different attributes from the others he had found. A lit candle would burn inside a container that was filled with common air, but in a container with only this gas, the candle burned with a flame many times brighter. It had other interesting characteristics. A mouse in an enclosed container filled with common air could survive for roughly fifteen minutes, but in this new air, a mouse could survive for an hour. He tried breathing this air himself and found it to be exhilarating. He thought there might be a future for it medicinally or even as something like a recreational drug.

Priestley continued to experiment with his new air. He inserted a

Joseph Priestley's work paved the way for Antoine Lavoisier's discovery of the true nature of fire.

lit candle into a vessel containing his dephlogistated air. As expected, its flame was extremely bright. The moment it went out, another lit candle was inserted, but it immediately went out. Priestley's interpretation was that the first candle had released its phlogiston, saturating the air with it. His dephlogistated air had become, in essence, phlogistated. When the second lit candle was introduced into this already phlogistated air, it had nowhere to go. Thus, it could not burn.

He repeated this procedure, but this time he did so with a living plant in the container. The first candle burned brightly and then went out, just as before. Another candle, inserted right after this, went out immediately. Oddly, when he waited ten days and inserted another lit candle, it again burned very brightly. It was as if the plant had replenished the air in some way. This was difficult to explain in terms of phlogiston. It was left for others, at a future time, to explain what was happening. It would be found that the first candle burned until the oxygen was depleted. The second candle went out because there was no oxygen to allow for combustion. The third candle, though, did burn. As a natural part of plant respiration, the plant, given enough time (in this case ten days), emitted enough oxygen so that the candle burned. Priestley would later share his discovery of this gas and its flammable nature with Antoine Lavoisier. It would be the missing link Lavoisier would need to upend the phlogiston theory and usher in a completely different theory.

Priestley, in the future, would face increasing opposition for his political and religious views. He was known as a dissenter—a term used for those that did not submit to the authority of the Church of England. Because of the conflicting beliefs, dissenters sometimes chose to simply leave England. This was the choice of the Puritans who boarded the Mayflower and sailed away. Joseph Priestley denied the divinity of Christ and rejected the concept of the Trinity. Rejection of the Trinity is the chief tenet of the Unitarian Church, and Joseph Priestley was integral in establishing Unitarianism in both England and the United States.[24] Already facing opposition for his religious teachings, he added contrary political views. He was in favor of the colonists' revolt in America. This was not a popular, though not unknown, viewpoint in England. He also became an outspoken supporter of the French Revolution. Although England did dabble with a monarch-less government under Oliver Cromwell, for most of its history, England was committed to its monarchy and had no intention of changing. Any support for France's overthrow of its king and queen seemed almost treasonous.

He published a document, *Reflections on the Present State of Free Inquiry in this Country,* in which he said, "We are, as it were, laying gunpowder, grain by grain, under the old building of war and superstition which a single spark may hereafter inflame, so as to produce an instantaneous explosion...."[25] This earned him the wrath of supporters of the royal government, and the nickname, "Gunpowder Joe." The French National Assembly chose to make Joseph an honorary French citizen which did not help his standing in England.

In Birmingham, where he was living at the time, he and some like-minded friends advertised and hosted a dinner party celebrating the second anniversary of the storming of the Bastille. To celebrate this was offensive to most Englanders for two main reasons; the visceral reaction to all things French, and the fact that the Bastille represented the first step in the taking down of a monarchy. This celebration led to what came to be known as the Birmingham Riots. It was a three-day affair in which dozens were killed and many buildings destroyed, including Priestley's church, house, and laboratory.[26]

He felt there was little choice but to leave, and did so, moving to London, then sailing for the United States in 1794. There, he peacefully spent the rest of his days, hobnobbing with many of the America's founding fathers and continuing his Unitarian preaching and teaching.

Lavoisier's Revolution in Chemistry

In 1771, French chemists, among them Antoine Lavoisier, had conducted experiments in the heating of diamonds. In one experiment, two enormous lenses, eight feet in diameter, mounted on a giant wooden structure, focused the sun's rays onto diamonds contained inside a glass vessel. The structure was located in front of what is now the Louvre Museum. It attracted a crowd of onlookers, either because of its giant size or simply because diamonds were involved. Two interesting things happened. When the diamonds were heated in a vacuumed container, no change took place. The diamonds grew hotter, but otherwise, were unaltered. However, when common air was in the vessel, the diamonds gradually disappeared. Lavoisier tried to duplicate this experiment to capture whatever gas might have been emitted, but his glass containers kept breaking due to the intense heat.

Guyton de Morveau, a lawyer that did chemistry on the side, had shown that metals gained weight when calcined. Other chemists had found this to be true as well, but this did not fit with the current theory. If phlogiston escaped, there should be a weight loss rather than a weight gain. Lavoisier had seen this himself in his own laboratory. In 1772, after burning phosphorus, what was left weighed more than the original. The same thing happened when he calcined sulfur. When it was heated, its surface formed a calx. The weight of the remaining sulfur plus the calx weighed more than the original sulfur. However, similarly to the case when diamonds were heated, no calx was formed if the sulfur was heated in a vacuum.

It was clear there was a weight gain, but unclear why it happened. He ran the experiment again with sulfur and with common air inside the vessel. When he opened the vessel, he noticed an audible inrush of air. Apparently, a vacuum or partial vacuum had been formed during the calcination process. It appeared that the common air inside the container, or perhaps a part of that air, he was not sure which, accounted for the weight gain. In further trials he found that the total gain in weight of the sulfur and the calx was the same as the weight of what was lost.

Much of Lavoisier's methodology was based on the concept that whenever change takes place, there was just as much matter before and after the change. There was just as much matter in an ice cube as there was in the water before freezing. There should be the same amount of matter after combustion or calcination as there was before. It was called the law of conservation of matter, although it has been, and in many places still is, called Lavoisier's Law. Others had proposed it, but Lavoisier had demonstrated its truth in his laboratory.

Lavoisier showed the conservation of matter to be true in case after case. Others had conducted experiments in which distilled water in an enclosed glass container was repeatedly boiled and allowed to cool. Eventually, there was a small amount of sediment that formed at the bottom of the container. This residue, plus the water it sat in, weighed more than the water before boiling. It seemed that the conservation of matter principle was being violated. To some, it seemed as if one of the four basic elements, water, could change into another, earth. What exactly was happening here was a mystery.

Lavoisier repeated those experiments. Using three pounds of rainwater, and after repeated boilings, he found that sediment did form, but also, the glass container weighed slightly less than before the experiment started. The amount lost in the glass was practically identical to the weight of the sediment. The sediment was not produced from the water but was a small amount of glass that had eroded during the repeated boiling. The original container, plus the original water, weighed the same as the container, water, and sediment after boiling.

For all practical purposes, the theory still holds today. However, since the advent of the nuclear age, it was found there are cases where it does not hold true. The theory is now better called the conservation of matter and energy. Albert Einstein showed that matter may be transformed into energy. Such is the case of nuclear fusion or radioactivity with his equation

$e = mc^2$ quantifying the amount. There is a loss of mass, but it is small. The first atomic bomb released a massive amount of energy, however the mass lost was approximately the same amount of that of a coin.

The proponents of the phlogiston theory had common sense on their side. When a material burned, it certainly appeared that there was less matter after the burning than before. That fit with the current thinking that phlogiston must have escaped. However, there was increasing evidence that what was left after burning weighed more than the original. Yet, to many there seemed no reason to discard the entire phlogiston theory. There were attempts to explain away the problem. Perhaps phlogiston did escape, but at the same time, impurities were introduced, thus accounting for the weight gain. It was even suggested that phlogiston somehow had negative weight.[27] Thus, the escaping substance of negative weight would cause what was left to weigh more. There was no explanation that was totally satisfying. Lavoisier thought something else entirely was going on.

Lavoisier felt that neither phlogiston, nor anything else, escaped when something was burned or calcined. To explain the weight gain, something must, in fact, be added. The only possibility seemed to be that it had come from the surrounding common air. Perhaps it was one of the recently isolated airs or a yet undiscovered air. In any event, phlogiston was not necessary. In fact, perhaps there was no such thing as phlogiston.

In 1773, Lavoisier told the Academy of Sciences that he felt these calces were simply a combination of air to the metal. Also, its reverse was "the disengagement of the air from the metallic calces."[28] Further, "I have even come to the point of doubting if what Stahl calls phlogiston exists."[29] He felt he was on the verge of a major transformation as he wrote, "The importance of the subject has engaged me to again take up all this work, which strikes me as made to occasion a revolution in physics and chemistry. I believe that I should not regard all that has been done before as anything other than indications."[30] While he was headed in the right direction, Lavoisier was still unclear as to what this unidentified air was. Was it common air as a whole or only a part of it, and if a part, what part?

Lavoisier had some idea of what Priestley and others were working on in England. John Hyacinth Magellan, a relation of the navigator Ferdinand Magellan, was working as a spy. He was employed by France's Bureau of Commerce, and his job was to find and send back information he thought might be helpful to French industry.[31] Up to this point, Lavoisier was not taking part in the investigation of various airs. New information from Magellan piqued his interest, however. While Magellan did supply some helpful material, a breakthrough would not take place until a particular dinner party.

In October of 1774, after his discovery of dephlogistated air, Joseph

Priestley went on a tour of Europe. One of his stops was in Paris where he was invited to a dinner hosted by Antoine and Marie Lavoisier. The Lavoisiers hosted dinner parties every Monday evening, often entertaining some of the leading lights of the age. Guests included those such as Joseph Lagrange; inventor James Watt; and the Americans Gouverneur Morris, Benjamin Franklin, and Thomas Jefferson. However, probably no dinner was as important as that attended by Joseph Priestley. There, he shared with the assembled guests the results of his recent experiments.

It was this dinner in 1774 that sent Lavoisier in the correct direction. Antoine, Marie, and the other guests listened as Priestley shared his news. He explained how he had placed mercury in a glass container, heating it until a powdery red calx was formed. He then heated that calx, with the result that the calx disappeared, the mercury was restored, and a gas emitted. This gas was his dephlogistated air (oxygen). He told the dinner guests of this gas' fascinating properties of greatly supporting combustion and seeming to have great health benefits. Lavoisier felt the common air, or a part of the common air, was responsible for combustion and calcination. However, he could not pin down what that air was. It seemed that Joseph Priestley had.

Lavoisier was not impressed with Joseph Priestley's methods. He felt them haphazard, often simply pouring random acids onto random metals to see what might happen. He felt his own experiments were more methodical and more precise. Lavoisier decided to rerun Priestley's experiment.

He heated a tube containing mercury until a calx formed on its surface. He found that the mercury and its calx gained weight, and this weight gain was the same as the weight lost from the air in the tube. Phlogiston did not escape. It seemed the calx was formed by part of the air joining with the mercury. Lavoisier had found that Priestley's dephlogistated air (soon to be called oxygen) combined with mercury to form the calx (now called mercuric oxide) and continued to do so until the oxygen was used up. When heated again, the reverse took place. Oxygen left the calx. When it did, the calx was restored to being mercury. The gas that was emitted was simply the same gas that had been absorbed in the initial reaction. The oxygen that joined the mercury to form the calx now left the calx restoring the mercury. In each step, the sum total of the vessel and its contents weighed the same. The weight gain that he, Morveau, and others had observed was not an overall gain. It was simply the shifting of the weight of the oxygen in the air to that of the calx.

Lavoisier came to find this is what happened in combustion as well. By his thinking, a burning object absorbs oxygen. Thus, by the principal of the conservation of matter, what remained after the fire should weigh more, as experiments showed that it did. That weight gain was due to oxygen being absorbed from the surrounding air.

His theory also explained the diamond experiment from a decade earlier. In a vacuum, the diamonds simply grew hotter, and when air was present, they shrank. It was not clear to the experimenters why there were different results depending on whether there was common air or a vacuum in the container. What was clear, or so they thought, was that the dissolving diamonds were again a sign of escaping phlogiston. Lavoisier found the solution. Without the presence of air, nothing happened because there was no oxygen to allow anything to happen. When air was in the vessel, changes took place. Diamonds are pure carbon. When heated, the carbon atoms attached to oxygen atoms and created carbon dioxide. The carbon leaving the diamonds is what caused the diamonds to shrink.

Lavoisier felt this new gas explained calcination and combustion. It explained the weight gain in substances and why it could not take place in a vacuum. It explained the disappearing diamonds. It explained everything.

Joseph Priestley, returning to England after his tour of Europe, made a presentation to the Royal Society in London detailing the results of his experiments. He described the results in terms of the phlogiston theory. Some months later, Lavoisier would also make a presentation, his to the French Academy of Sciences. He detailed his latest experiments—the calcination of mercury, the calx that formed, and then being able to reverse the process. He made no mention of Priestley anywhere in his paper. Clearly, Lavoisier's experimentation was based on the work of Priestley, and while his conclusions were different, he did not begin to travel that path until their dinner together. Priestley expressed his understandable dismay. He thought his previous work should have been recognized by Lavoisier. He did not agree with Lavoisier's scientific conclusions and remained a believer in phlogiston for the rest of his life, but right or wrong, he deserved acknowledgment for his prior work.

Lavoisier had shared his results and given credit to fellow scientists in the past, so it is difficult to see why he did not do so in this situation. Perhaps it was because he disdained Priestley's methodology. Perhaps it was another example of the French-Anglo hostility, or that it was simply an oversight. Regardless, it is a blemish on Lavoisier's career.

Through the years, Lavoisier would continue to develop his theory, and many would disagree with him. The issue came to a head in 1787 when he made a presentation to the Academy that was, to say the least, tense. In fact, the response to his presentation was described as chaos.[32] He was interrupted repeatedly with shouts from his fellow scientists. Many had logical objections to the new theory. Many were simply angry about their theoretical world being turned upside down. Prior to any thought that Lavoisier had about debunking phlogiston, Condorcet stated, "If ever there was anything established in chemistry, it is surely the theory of phlogiston."[33] In a letter written by Benjamin Franklin, "I would like to hear how M[onsieur] Lavoisier's doctrine supports itself, as I suppose it will be controverted."[34] Franklin was quite right about that. Scientists do not give up their theories easily. While eventually virtually everyone (except Priestley) would move from phlogiston to this new theory of oxidation, it took years until there full was agreement.

As he developed his theory, Lavoisier worked in related areas. In 1782, he teamed with the mathematician Pierre Laplace to explore respiration. They would ultimately show that the process of respiration is essentially a combustion reaction. They found that oxygen is taken from the air, carbon from food, and in a chemical reaction, there is a release of carbon dioxide and energy.[35]

Lavoisier and Laplace had developed and used the first calorimeter—a device used to measure the amount of heat given off by a chemical reaction.[36] The calorimeter was filled with ice, with the volume of water produced by the melting ice being an indicator of the amount of heat an activity gave off. Literal guinea pigs were used in their experiments. However, the real hero was their figurative guinea pig, Armand Séguin. He was a chemist who later became a rich man thanks to a process he developed for tanning leather. He would be paid a great deal to outfit French armies during the upcoming Napoléonic Wars. For now, the willing volunteer offered up his body to be used to measure the effects various activities had on respiration and on heartbeat. They examined behaviors such as resting, exercising, and eating. They also examined the effects of being in a cold room versus a hot one. Séguin was sewn into a form-fitting silk bag. The bag was weighed, then Séguin was weighed before and after any of the experiments. He breathed through a tube and, at times, in through one tube and out through a different one.[37] Although their experiments were cut short due to the onset of revolutionary turmoil, the chemists gained valuable information about the effects these various activities had on the body. Some discoveries might have seemed obvious—more oxygen was used when exercising than at rest. Some were not—the act of digesting food caused an increase in the amount of oxygen used.

Priestley had shown that his dephlogistated air (oxygen) was used up when a candle burned. Lavoisier went farther. He placed a lit candle in an enclosed jar filled with common air and placed the whole structure under water. When the jar was opened, water filled one-fifth of the jar, showing that this new gas made up one-fifth of the atmosphere.[38] (The actual amount is 20.9 percent.)

A link was found between Joseph Priestley's gas and water. Henry Cavendish, the discoverer of hydrogen, was able to create water. He found that mixing hydrogen and common air inside a vessel, and introducing a spark, caused a small amount of water to form on the inside wall of the vessel. This seemed odd. How do these invisible, ephemeral gasses come to make up this most important of substances? Yet, it seemed to be the case. His explanation for what was taking place, though, was off the mark. For Cavendish, water was not a compound but a basic element of nature. He theorized gasses contained water, and when the spark was introduced, the water was released. He was incorrect and compounded the error by explaining the results in terms of phlogiston.

Lavoisier ran his own experiment in which a mixture of hydrogen and oxygen and a spark again formed water. The total mass of water formed was the same as the mass of the hydrogen and oxygen that was used up. He stated that the basic elements were hydrogen and oxygen, not water. He then completed the process by breaking down water and again obtaining the original gasses, hydrogen, and oxygen. He announced to the Academy of Sciences, "Water is not a simple substance, it is composed weight for weight of inflammable air [hydrogen] and vital air [oxygen]."[39]

Lavoisier had now taken three of the four original Aristotelian elements and shown them not to be elements at all. With some assistance from others, he had demonstrated that fire is a process of oxidation, common air is a mixture of gasses, and water is a compound. And the fourth element, earth, was looked on differently as well. There were metals that were thought to be compounds, such as phosphorus, because they contained phlogiston. When its phlogiston escaped, the calx remained. Thus, the calx and phlogiston were the elements, and the phosphorus was a compound of them. Lavoisier established this was not the case and thus showed that many substances that were thought to be compounds were actually elements.

A New Vocabulary

Alchemists, in their process of searching for the philosopher's stone and magic elixirs, also discovered many substances. Upon their discovery they were given names, but often this led to more confusion than clarity. There were no standards for how names were assigned. A substance might be named for its discoverer, its appearance, its place of origin, a mythic being, or even given a name that just sounded interesting. Different countries or occupations might adopt their own names for a substance, with the result that a single substance could have multiple names. As Lavoisier later wrote, "…as ideas are preserved and communicated by means of words, it necessarily follows that we cannot improve the language of any science without at the same time improving the science itself; neither can we, on the other hand, improve a science without improving the language or nomenclature which belongs to it."[40]

Chemists were dealing with a jumble of completely unhelpful names. Lavoisier worked with three other French chemists for six months to completely rename 700 substances, creating a vocabulary that was a much cleaner, more efficient system.[41] In 1787, they published the results of their work in the book *Méthode de Nomenclature Chimique (Method of Chemical Nomenclature)*. Because of their efforts, in the future, the spirit of salt would be known as hydrochloric acid. Rather than Saturn sugar, there would be lead acetate; instead of shrimp eyes salt, calcium acetate. Philosophic wool became zinc oxide. The new names gave information about the substance. The chemical HCl is called hydrogen chloride and each molecule is made up of one hydrogen and one chlorine atom, the chemical symbol NO is nitrogen oxide, and BrCl is bromine chloride. Suffixes such as -ic, -ide, and -ate gave future students and chemists information on what they were working with. For example, if an atom gains an electron, its state becomes that of a negative ion. Chemists can denote this with the suffix -ide. Thus, the negative ions of fluorine, bromine, and chlorine, are called, respectively, fluoride, bromide, and chloride.

There was resistance to this new language. Thomas Jefferson felt the effort, "premature, insufficient, and false … upon the whole I think the new nomenclature will be rejected after doing more harm than good."[42] Joseph Black clearly was not in favor when he said, "When changes are thus made in the names of things which are familiar to us, I believe most people find them disgusting at first, on account of the shock and derangement which they give to the habits we had formed before. These latinized French words appeared to me at first very harsh and disagreeable."[43] Yet, the four chemists would prevail. The three that joined Antoine Lavoisier were Antoine Fourcroy, Claude Berthollet, and Guyton de Morveau. They became known jointly by the name "The French Chemists."

Antoine Fourcroy was born in 1755. He was educated in Paris in the field of medicine, but shifted to the study of chemistry, becoming a professor at the Medical School of Paris. From 1802 to his death in 1809, he was the Minister of Public Instruction in France. In the midst of the revolution, he served in the National Convention. He became president of the radical Jacobin party and was a member of the Committee of Public Safety. Historians are divided on his role in the death of Lavoisier. Some have said Fourcroy did his best to speak in his defense. Others have said that he silently consented to his death or even spoke against him.

Claude Berthollet was born in 1748 on the eastern border of France in the city of Talloires. He rivaled Lavoisier for his contributions to French society. Like Fourcroy, he obtained a medical degree, his from the University of Turin. He applied his knowledge of chemistry to the advancement of French industry in a variety of areas, such as in the production of steel, dyes, and gunpowder. He isolated ammonia and found many of its properties. He was the first to demonstrate the bleaching action of chlorine gas.[44] He worked with Gaspard Monge to establish the École Polytechnique, where he would teach chemistry. A favorite of Napoléon Bonaparte, he sent Berthollet to critique and then confiscate works of art after his victorious Italian campaign. He was appointed to important positions, such as the committee in charge of the mint and the committee that would establish the metric system. He was among the scientists who accompanied Napoléon to Egypt. Berthollet tried to avoid the violence of the revolution by moving to Aulnay, almost 300 miles away from his home. He was ultimately brought back to Paris and appointed to the senate by Napoléon. He would later be in favor of Napoléon's banishment to St. Helena following the defeat at Waterloo.

Guyton de Morveau was supportive of the French Revolution. At various times he was a member of the Legislative Assembly, the National Assembly, and the Council of Five Hundred. He voted in favor of Louis' execution. He was on the Committee of Public Safety, but only there for three months. He resigned to assist the French army's efforts, manufacturing firearms and helping to form the nascent science of military ballooning.

Two years later, after publishing *The Method of Chemical Nomenclature*, Lavoisier published what has been called the first chemistry textbook.[45] It was titled *Traité Élementaire de Chemie* (*Elementary Treatise of Chemistry*). It was thorough and clear and did much to advance the chemistry of the day. For many years it would be the chemistry textbook used in classrooms throughout the world. In it, he summarized a great deal of what he and others had been working on. He covered his conservation of matter principle in which he famously states, "Nothing is lost, nothing is created...." He covered the nature of gasses and presented his case against the theory of phlogiston. He explained the make-up of water. He stated that substances can exist in three states—gas, liquid, and solid; and the changing states are only physical changes. The particles that make up the substances simply get closer or farther apart. He defined an element as a single substance that cannot be broken down further, and that all chemical substances were elements or made up of elements. In his book, he listed 33 such elements, although some in his list were later found to be compounds, and he stated the existence of two, light and caloric, that were not substances at all. He gave oxygen its current name, doing away with Priestley's dephlogistated air and Schleele's fire air. For a time, he called it "highly respirable air," which actually would have been a more accurate name. He took the name oxygen from Greek roots (literally, acid generator). Oxygen is a part of many acids, but Berthollet soon discovered that there are acids, such as hydrochloric acid, that are formed without oxygen. Regardless, the name stuck. He also gave the name hydrogen to the gas Boyle discovered, replacing the name inflammable air. He chose that name because it was contained in water—hydro-generated. Those names lasted, although his naming of nitrogen as "azote" did not. The text contained excellent, detailed drawings, all done by his wife, Marie. The book was translated into English the following year and other languages thereafter. It taught chemistry to future generations.

To Benefit Mankind

It is difficult to imagine that, with all his work in the field of chemistry, Lavoisier had time for anything else. Yet, he spent perhaps as much time in his work assisting the French government as he did in the laboratory. Because of his abilities and willingness to do so, he was called on by the monarchy, and later the revolutionary government, to take on many projects and positions.

Lavoisier was asked to be a member of, and often the head of, a great number of important committees. He was the head of a committee that was asked to standardize the weights and measures of France. It did that and more, leading to the development of the metric system. In 1793, as a member of the Commission of Arts and Professions, he presented a report which was the commission's plan for a system of free public education. In 1785 he became the head of the Academy of Sciences. He was among the first to join The Society of the Friends of Black People, an organization whose goal was to abolish the slave trade and ultimately all slavery throughout France.

The king appointed Lavoisier to be the head of an agricultural committee. He made a purchase of land from General Rochambeau that he used to further agricultural science.[46] Rochambeau was preparing to leave for America to lead French troops in the American War of Independence. He, Lafayette, and Washington would soon trap Cornwallis at Yorktown, effectively ending the war. After returning to France, Rochambeau was arrested during the Reign of Terror. It was only with Robespierre's death and the end of the Reign of Terror that his life was spared from the guillotine. There were many, such as Rochambeau, that were arrested but avoided execution only because of the timely death of Robespierre and the subsequent demise of the Committee of Public Safety.

Lavoisier used the land he purchased from Rochambeau to build a model farm. He researched agricultural methods and worked with local farmers to develop efficient practices. He demonstrated the use of crop rotation. His knowledge was such that George Washington sought his advice on farming his own property at Mount Vernon.[47] He convinced the government to set up a fund that area farmers could invest in as a hedge against lean years, and, from his own funds, he made loans of money to those in need. Many children, especially among the poor, were not able to attend schools, so Lavoisier began a school for the children of local farm workers.

He continued to write reports that sought to assist with society's issues. He proposed improvements to the nation's prisons and hospitals. Prisoners had to pay the expenses they incurred while being incarcerated. The prison cells were so crowded, roughly one square meter per inmate, that it often meant contorting one's body on the floor in order to find a place to sleep.[48] The prisons had a lack of sanitation, a lack of ventilation, and an overabundance of vermin. Hospitals were not necessarily much better. Lavoisier proposed routines of washing clothing on a regular basis, disinfecting areas, and proper ventilation.

The Seine River, which runs through the heart of Paris, was a primary source of both drinking water and sewage disposal. He reported on his plan for increased sanitation and drinkability of its waters. He also wrote reports on soap making, coal mining, and water storage on ships. Many of his proposals ended up not being acted upon, either due to a lack of vision by those in power or simply due to the increasing turmoil of the times.

Perhaps his most far-reaching appointment was to the Gunpowder and Saltpeter Administration. Saltpeter is a prime ingredient in gunpowder, but it was in short supply and of inferior quality. One reason for the shortage was the amount of ammunition that was used during recent wars. The Seven Years' War, especially, drained France's reserves. Also, in the peace treaty ending the war, France lost its colony in India, a primary source of saltpeter. Lavoisier fixed both the problems of quantity and quality. Instituting more efficient methods of production, establishing training programs, building new factories, and even sending individuals to India to investigate why it held such reserves were some of his innovations. He developed chemical processes that dramatically improved the quality of the gunpowder. French cannonballs could fly fifty percent farther than they had in the past.[49] A major reason for Cornwallis' surrender at Yorktown was that colonial and French cannons could reach the British lines, while the reverse was not true. In three years' time, he doubled the amount of gunpowder that France possessed. There soon was enough gunpowder available so that the surplus could be sold to other countries, enabling France to supply the American colonies, which was instrumental in their defeat of England.[50] Lavoisier later stated that "One can truly say that North America owes its independence to French gunpowder."[51]

The production of gunpowder was formerly run by a private company. In 1775, the comptroller of France at the time, Anne Robert Jacques Turgot, turned the Gunpowder Committee into a government agency and named Lavoisier as its head. A requirement of holding that position was that Lavoisier live at the arsenal, which was a storage location for Paris' gunpowder. While that prospect perhaps did not seem initially inviting, the arsenal contained a large apartment which Antoine and Marie made to be quite elegant.

With his own money, Lavoisier purchased equipment that could be used in an adjoining laboratory. It became perhaps the best equipped laboratory in the world at that time. Much of his scientific success was his ability to make precise measurements. That was, in part, because he could afford the finest equipment and, in part, due to the meticulous habits he employed to obtain precise measurements. He had an enormous balance made that was three feet in length and had an accuracy within 0.00025 percent.[52] He regularly took calculations out to five decimal places. The care he took in his experimentation came to influence the way many of the scientists in the future conducted their own experiments.

As fitting with his nature, Lavoisier had scheduled his days precisely to maximize his efforts. He rose at five o'clock each morning, working in his laboratory from six to nine in the morning. The rest of his morning and until six o'clock in the evening was spent in administrative work involving his General Farm and Gunpowder Committee duties.[53] After dinner, he spent the evening back in his laboratory until ten o'clock. On Sundays he spent the entire day in the lab doing experiments with his specially selected students. According to Marie, it was his favorite day of the week.

The arsenal was located next to the Bastille prison. On July 14, 1781, doubtless Antoine and Marie could look out from their apartment and see the crowd forming, culminating in the storming of the fort, the release of the prisoners, and the macabre celebration that followed. There appears to have been no direct negative impact on the Lavoisiers that day. Days later, there was a shipment of ammunition delivered to the arsenal. There was nothing odd about that, except rumors were spreading that the ammunition was intended to be sent on to Austria, who was threatening war with France. An angry crowd descended on the arsenal. Lavoisier was ultimately able to calm them and explain the situation, but for a time, the fate of the arsenal and of Lavoisier was in doubt.

One of the roles of the Academy of Sciences was to test the validity of scientific claims. One reason for doing so was to safeguard the public from being taken in by charlatans. Also, before such a concept as copyright law existed, the Academy served to establish proper credit for a discovery or invention. Such was the intent when the Academy sent a team of its members to investigate the claims of flight by the Montgolfier brothers. Another case involved an individual who said he had a method with which he could find sources of underground water, and thus, apt locations for digging wells. The Academy found this claim to be unspectacular, feeling the water table was such that a well dug deep enough anywhere would likely find water at some point.

A more difficult situation was that of Franz Anton Mesmer. Mesmer was born in 1734 in what is now Germany. He obtained a medical degree from the University of Vienna, though apparently his education was not as thorough as one would have hoped. Mesmer stated that gravitational forces of the celestial bodies could have an effect on the human body, negatively impacting a person's health. However, a trained individual (such as himself) could undo any negative effects through the use of magnetism.[54] He might use what he claimed was a magnetized wand, or perhaps pour magnetized water over one of his patients. Today, to be mesmerized has a slightly different meaning, but stems from this original approach by Dr. Mesmer. Through flawed, his work later influenced Sigmund Freud and led to the science of hypnotism. His approach was discredited while he was in Vienna, so he moved to Paris in 1778 and there found a larger, more receptive following.

A committee, which included Lavoisier, Jean-Sylvain Bailly, Benjamin Franklin, and the unfairly infamous Joseph-Ignace Guillotin, was selected to investigate this phenomenon.[55] In 1784, the committee reported to the Academy, and not surprisingly, did not find his methods to have medical validity. Their opinion was that there were, in fact, cases where individuals did show an improvement in health, but those seemed to be more due to a belief in the doctor than the treatment itself.[56]

The most troubling investigation, and the one with the greatest impact on Lavoisier's future, was that of Jean-Paul Marat. Born in 1743, in what became the country of Switzerland, Marat left home at 16 and went to France. He became a leading radical voice of the revolution. His publication, *L'Ami du Peuple* (*The Friend of the People*), made him an important voice of the Jacobin party. Marat became a member of the National Convention and later was a leader of the Committee of Surveillance, an arm of the Committee of Public Safety. The Committee of Surveillance did what their name implied. It included fellow revolutionaries Georges Danton and painter Jacques-Louis David.

Another interest of Marat's was science. He joined the growing field of those exploring the nature of fire. He was not a proponent of phlogiston or Lavoisier's theory of oxidation, but of a new, although incorrect, theory of his own making. He also wrote about the nature of light, unwisely disagreeing with Isaac Newton regarding his theory of how light could be broken down into various colors.

Marat asked the Academy to investigate his claims. The Academy did so, approving his methodology, yet coming short of endorsing his results. Later, when writing of his scientific work, Marat claimed that the Academy did approve of his results. Lavoisier took exception to this. He felt an unqualified endorsement was well beyond anything the Academy had intended and insisted the record be set straight. The Academy then did, in fact, make this correction and stated they did not necessarily endorse Marat's results. This, of course, was personally embarrassing to Marat. He never forgave what the Academy, and specifically Antoine Lavoisier, did to him.

In September of 1789, in *L'Ami du Peuple*, Marat made it clear to the world how he felt about his nemesis. "Lavoisier, putative father of all noisy discoveries; he has no idea of his own so he appropriates those of others; but since he cannot understand them he abandons them again as easily as he adopts them, changing systems as he does his shoes. In a space of six months he has picked up in turn the doctrines of fire, igneous fluid, latent heat. In shorter spaces I have seen him first infatuated with pure phlogiston then ruthlessly denouncing it. Some time ago, following the lead of Cavendish, he discovered the secret of making water from water. Then, imagining that this fluid is composed of pure air and inflammable air, he changed it into combustibles. If you ask me what he has done to warrant such praise, my reply is that he has got for himself an income of one hundred thousand livres, has placed Paris in prison with his great wall.... Proud of his achievements, he rests on his laurels while his parasitic followers praise him to the skies."[57] Marat would put forth many such diatribes, all with the goal of destroying Lavoisier's reputation.

As unrest grew, many went into hiding, sought to flee the country, or simply took on a lower profile. In August 1792, Lavoisier resigned his government positions, moved out of his living quarters at the arsenal, and took up residence at his farm, a hundred miles from Paris. A father and his son were left in charge at the arsenal. Days later, it was taken over by the National Guard. The father and son were put into prison where the elder committed suicide.[58] Their arrests were part of a sweep of the entire city that was taking place at that time. Overseen by Marat, the city gates of the wall surrounding Paris were closed, and thousands of supposed enemies of the state were rounded up. The arrests were followed by the September Massacres in which many of those recently locked up would die in their cells.

Antoine and Marie considered leaving the country. They were invited by friend and government economist Pierre Samuel du Pont to escape with him to America. They chose not to, in part, because of a commitment he felt to the Academy, and the danger it was in by those that were now running France. The fact that the Academy was originally established by royal decree and funded by the king automatically put it under suspicion, and Marat's paranoid rantings did not help its reputation. He referred to its members as "automatons" and "vain men,"[59] while Jacques-Louis David said they were "incompatible with the reign of liberty."[60] Lavoisier returned to Paris in November to try to save the institution. When the government stopped payments to its members, Lavoisier paid them out of his own pocket. His efforts would not be enough to save it, and it was closed by order of the National Assembly in 1793.

By June of 1793, many members of the Girondist party had been arrested or had fled the country. The Jacobins were now in firm control, led by Marat, Danton, and Robespierre. While bathing in his home, Marat was stabbed to death by 24-year-old Charlotte Corday, a Girondin sympathizer. Four days later, she would go to the guillotine for her act. Marat was memorialized by fellow Jacobin Jacques-Louis David in his famous painting, *The Death of Marat*. While Corday's objective was the defeat of the Jacobin movement, it had quite the opposite effect. Marat was a political leader and a celebrity, and in the wake of his death his popularity soared. It led directly to the Reign of Terror. The Law of Suspects allowed for the arrests of so-called supporters of tyranny and enemies of the people. In the eyes of those in power, that description fit the supervisors of the General Farm.

Financial Crises

The General Farm, which employed thousands to collect the taxes on various goods and services, was managed by an association of a few dozen partners, with some 300,000 doing the actual collection of the taxes.[61] When he was 24 years old, Lavoisier became one of those partners when he bought a one-third share, which he was later able to turn into a full share.[62] The partners were in charge of collecting the tax money, taking out their share, then transferring the remainder to the royal government. For these tax collectors to make any money, it was necessary to collect more than was owed. Unscrupulous collectors collected a lot more. There is no evidence the Lavoisier was of that sort.

From the Bible's Matthew to the present day, being a tax collector has never been the way to achieve popularity. Present-day tax collectors do not face the vitriol of their predecessors. Having money taken out of pay-checks or sending a check to a faceless government agency has reduced the personal nature of collections. In the years leading up to the French Rev-olution, the fact that high taxes were being collected at a time of financial difficulties simply added to the animosity.

The reason for the lack of money went back a hundred years. There had been years of extremely low temperatures, drought, and freak storms causing not just a lack of food for the people, but also a lack of tax money for the government. The government's response was to raise taxes. Although the natural disasters were beyond anyone's control, other factors were very much in the king's control. France fought directly in, or finan-cially supported, several wars. This was not only expensive, but sometimes resulted in the loss of revenue-producing colonies.

Additionally, there was overspending by the royalty. It was a focal point of complaints by the people, though its effect on the financial defi-cit was not nearly as great as that of wars and weather. Beginning with Louis XIV, Paris gained much of its splendor from major building proj-ects, such as the opulent Versailles Palace, just outside the city. The results were beautiful, but expensive. Louis XV and XVI, their families and courts, were perhaps not quite as extravagant as the Sun King, however they still spent lavishly on parties, clothing, and other frivolities. While Marie Antoinette was never recorded as suggesting that if there was no bread, the poor could eat cake, the people felt it was the kind of thing the uncaring queen might very well say. Among the people, she acquired the name Madame Deficit. There is no question that Marie, and almost every-one associated with the royal family, spent extravagantly. However, while Marie was made out to be a selfish, uncaring woman, many that knew her spoke of what a kind person the queen was and how much she cared about the people of France. An example of her kindness can be found in the last words she spoke. Accidentally stepping on her executioner's foot, she told him, "Monsieur, I beg your pardon. I did not do it on purpose."[63]

It fell to the government comptrollers to keep France solvent. A series of comptrollers; Necker, Calonne, Turgot, Colbert, all tried, but ultimately failed to right the ship. Though they might strongly advise against it, they had little control over royal spending on gowns, palaces, parties, or wars. It was the king that ultimately made the financial decisions. It was the comp-troller's job to find the money. Money, though, was not easy to find. The first and second estates, the clergy and nobility, were largely exempt from taxes. That left the third estate to foot the bill. They were most of the coun-try's population, but financially they ranged from poor to middle class.

There were abundant taxes. Among them were custom duties on goods coming into Paris and at various other points in the country. There were almost 2,000 tollgates throughout the country used to collect that money.[64] Those tollgates were strategically located on roads entering France and on roads entering cities. It was the job of the General Farm to collect the taxes. Beside the physical collection of money, they were to pursue those that smuggled goods, in an attempt to avoid paying taxes. Lavoisier began his career with the Farm as a regional inspector for its tobacco commission and traveled the state in fulfillment of his duties. The commission's job was to find contraband tobacco. Sometimes dealers surreptitiously added ash to increase their profits, though Lavoisier would soon design a chemical test to tell if ash had been added to a sample.[65]

For the tobacco that did not evade the tax, a controversy developed over the watering of the product. Some watering was done to begin its fermentation.[66] The question was how much.[66] Those that grew the tobacco complained that too much water increased the overall weight unnecessarily, which meant paying more in tax and cutting into their profits. Too much water could also ruin the product. Lavoisier was one of the General Farmers that were in favor of keeping the watering of tobacco to a minimum. Allegations regarding the watering of tobacco would be held against the tax collectors at their trial.

To avoid paying taxes, the smuggling of goods into Paris was becoming rampant. Lavoisier figured that a fifth of the goods entering the city did so illegally.[67] There were stations on the roadways of France where those transporting the tobacco were stopped and assessed the duty. Smugglers, however, found creative ways to avoid them. Lavoisier had been appointed to the General Farm's Central Administrative Committee in 1783. In that role, his proposal, which was accepted, was that a wall be built around the entire city of Paris. It would stand six feet tall with sixty-six tollgates, several of which are still in existence today.[68] The wall was not at all popular. Smugglers certainly did not appreciate it, but the citizens of Paris did not either. They said they felt imprisoned by living in a walled-in city. Lavoisier did not give final approval for the wall, nor was he in charge of its construction, but much of the blame from the populace went to him. The construction of the wall fell to the leading French architect of the day, the visionary Claude-Nicolas Ledoux. Because of royalist connections, many of his projects would be vandalized and torn down in coming years. These connections also led to his arrest at the height of the revolution, though he was later released. Following the storming of the Bastille, the mob tore down this hated wall.

Arrest and Trial

In spite of all that Antoine Lavoisier had done for the good of France, many in the government, led by Marat, were building a case against him.

Lavoisier was no longer a part of tax collection, having resigned all of his government positions months before, but his activities there and other past events worked against him. His work as a tax collector with the General Farm and the suspicions surrounding the Gunpowder Commission, not to mention Marat's vitriolic attacks, kept his name on the list of those to watch. He supported the king and was in favor of a constitutional monarchy, but, as much as he could, he sought to stay away from politics. It was a middle ground, but a middle ground was generally not appreciated by the Jacobins.

On November 24, 1793, a warrant was issued for the arrest of Antoine Lavoisier, his father-in-law Jacques Paulze, and others associated with the General Farm.[69] Paulze was 70 years old at the time of his arrest and would go to the guillotine on his next birthday. For the next six months they were held in Port-Libre prison, a former abbey that had recently been converted to a prison due to the need to house the many recently arrested. He and his father-in-law would share a cell. Although a handful of prisoners escaped execution by bribing officials or using their influential friends, nearly thirty of the former tax collectors would go to the guillotine.

Marie was now all alone. The livestock and other goods at the farm were confiscated and the house placed under guard. The property was soon sold at auction. (In the future, the farm with its mansion would be in private ownership until being donated to the church when it became a mental hospital. It is now a bed and breakfast.) Almost all the furniture in his and Marie's Parisian apartment was removed with only a bed, three chairs, and a vase being left.[70] His laboratory equipment was taken. For the time being, Marie was allowed to stay in their apartment with its meager furnishings. In the weeks ahead, she would do what she could to obtain Antoine's, her father's, and others' release. While in prison, Lavoisier was able to write two more volumes regarding his research in chemistry. Later, Marie would see to it that they were published. On June 14, 1793, more than a month after her husband's execution, Marie was arrested by the Committee of General Security. It was not clear then, nor now, why. She spent 65 days in jail before the Committee decided to release her.[71] Through her own and others' campaigns, she eventually regained the property and almost all the items that had been confiscated. She would live in Paris the rest of her life, still hosting parties and being a focal point of French society. Years later she married Benjamin Thompson, an English physicist. It was a stormy marriage ending in a divorce after only four years. Marie passed away in 1836 at the age of 78.

Meanwhile, those arrested waited in jail while the state constructed its case. The primary investigator for the trials was Antoine Dupin. On May 5, 1794, he submitted a report of 187 pages, detailing the state's charges against the supervisors of the General Farm.[72] They were accused of making illicit gains as they performed their duties. It was claimed they had cheated the state out of money it was owed and used the treasury as its own personal bank account. For good measure, the old charge of watering down the tobacco was thrown in as well. On May 6, the prisoners were transferred to another prison, the Conciergerie, to await their fate. Two days later, the evening before their trial, the prisoners were finally told of the charges against them.

On May 8, 1794, they faced the Revolutionary Tribunal. The trial lasted only three hours. In fact, the trial, conviction, sentencing, and execution all took place the same day. The state presented its case against these "vampires."[73] There were three judges and a seven-person jury, the leading judge being Jean-Baptiste Coffinhal. The prosecutor, Antoine Quentin Fouquier de Tinville, made his presentation. Attempting to unravel finances going back several years proved difficult for the defendants. At one point, accused of stalling for time, they were told answers to questions must consist of only a "yes" or a "no."[74]

There were attempts by associates of Lavoisier to show his innocence and his value to France. They had no impact. Legend has it that Coffinhal at one point replied, "The Republic has no need of scientists."[75] While that perhaps was not actually said, it appeared that France at least felt it had no need of Lavoisier. Only three months later, during the Thermidorian Reaction, Coffinhal would himself be tried, convicted, and sent to the guillotine. The life of the prosecutor, Antoine Quentin Fouquier de Tinville, would end the same way.[76]

At four o'clock in the afternoon, with hands tied behind their backs, the prisoners were loaded into carts for the short ride from the Conciergerie to the Place de la Révolution. It was the same mode and route taken by Marie Antoinette six months before. They would be at the same guillotine, and with the same executioner as that of Louis XVI, Charlotte Corday, and hundreds of others in the past months. There was roughly one execution per minute. The prisoners waited their turns. Jacques Paulze was third in line and Antoine Lavoisier was fourth. After the executions, 28 lifeless bodies were loaded onto the wagons that had brought them there, their heads collected in a large basket.[77] Their bodies were taken to the Cemetery of Errancis, located outside of Paris, where they joined almost a thousand others that had suffered a similar fate. An apartment complex now stands on that spot.

Though perhaps apocryphal, Joseph-Louis Lagrange is said to have commented regarding Lavoisier's death, "It took them only an instant to cut off that head, and a hundred years may not produce another like it."[78] The loss of Antoine Lavoisier is almost incalculable. He died at the age of fifty and certainly had many years of research and benevolent work left to him. The French Revolution hampered a good deal of his work while he was alive, then ended his life early.

More Chemists and More Chaos

Antoine Lavoisier was only one of many French scientists who suffered from the effects of the French chaos of those days.

Philippe Friedrich Dietrich was a member of the Academy of Sciences. He worked with Italian scientist Alessandro Volta (of whom volt and voltage are named). He authored a three-volume work on the use of ores by French industry. During the years of the revolution, he was the mayor of the city of Strasbourg. As a patriotic gesture, he commissioned the writing of what would become the French national anthem, *La Marseillaise*, which was first performed in his home. He spoke out against the violence of the revolution and in support of priests who had refused to take loyalty oaths to the new government. Robespierre had called him a dangerous man, and the Revolutionary Tribunal sentenced him to the guillotine in 1793.[79]

Jean-Baptiste-Gaspard Bochart de Saron was a mathematician and astronomer. He possessed a large number of telescopes that he used and freely loaned to other astronomers. William Herschel had discovered what he believed to be a new comet. However, calculations by de Saron showed the orbit was not what Herschel thought it to be, which lead to its ultimately being declared the planet Uranus. He became president of the Parlement of Paris (a court of law), although he generally tried to stay away from the mounting political turmoil. In 1793, in an unusual ordering of justice, former members of parlement were first arrested, then evidence against them was sought. After a brief trial in which there were no witnesses, de Saron and fellow associates were sentenced to the guillotine.[80]

Chrétien de Malesherbes was a royal administrator and a member of the defense team at the trial of Louis XVI. He was instrumental in getting Diderot's *Encyclopédie* published and was a well-respected botanist and a member of the Academy of Sciences. He was arrested for treason in December of 1793, primarily for his defense of Louis XVI at his trial. He, his daughter, and his granddaughter were sent to the guillotine the same day, the following April.

Louis Alexandre de La Rochefoucauld studied the natural sciences and was a member of the Academy of Sciences and the Royal Society of Medicine. In 1789, he was a member of the Estates General and supported the desires of the third estate as events escalated toward the Tennis Court

Oath. As things began to heat up, La Rochefoucauld decided it was time for him and his family to leave France. Those nobles seeking to leave France were seen as unpatriotic, if not traitors. Before making it out of the country, he was arrested and executed in September 1792.[81]

There were others that attempted to leave the country. Some were successful and some were not. An unknown number suffered untimely deaths simply due to stress. Some took their own lives rather than face the Tribunal. The du Ponts were an example of a family that survived. They survived, but their departure was a great loss to the nation of France.

Pierre Samuel du Pont de Nemours was born in Paris in 1739 and became an important French economist. France's comptroller, Turgot, selected him to be inspector general of commerce. Although he was serving in the royal government, he supported the revolution and was president of the National Constituent Assembly. Though supportive of the revolution, he was among those that were there to defend Louis and Marie when a mob came for them in the Tuileries. He was arrested in July of 1794 and condemned to death, but Robespierre's death and the end of the Reign of Terror took place before the sentence was carried out. After his house was ransacked, Pierre decided it was time to leave, and in 1799, he brought his family to America. Years later, Pierre returned to France at the request of President Thomas Jefferson to assist with the negotiations that would lead to the Louisiana Purchase.[82] Otherwise, his new home would be the United States.

The du Ponts were good friends with Antoine and Marie Lavoisier. Pierre had proposed marriage to Marie after Antoine's death but was rejected by her. He had three children by an earlier marriage, the youngest being Éleuthère Irénée, who was 28 at the time of the departure to America. Éleuthère was a protégé of Lavoisier and had learned a great deal about gunpowder production from him. They had worked together as members of the gunpowder committee as well as other ventures. In America, he put that knowledge to use and started his own business, initially only producing gunpowder, but later expanding. Originally, he was going to name his company Lavoisier Mills after his mentor, but ultimately decided to call it E.I. du Pont de Nemours & Company. His company stayed in the family, and the chemical company known as DuPont would become one of the largest and most successful in the world for the next 200 years.

His company would one day adopt the motto, "Better Living Through Chemistry." So much more for society and for chemistry could have been accomplished if Antoine Lavoisier's, and many other chemists' lives, had not been cut short.

IV

The Measurement of All Things

"Never has anything grander and simpler and more coher-
ent in all its parts come from the hands of men."
—*Antoine Lavoisier[1]*

The Magna Carta states, "Let there be standard measures of wine, ale,
and corn throughout the kingdom." The Bible says, "You shall do no wrong
in judgement, in measures of length, or weight or quantity." The French
Encyclodédie spoke to the issue of needing fair measures in the country.
In a France facing revolutionary change, the lack of measurement standards was on the list of grievances presented by the people to the Estates
General.

Buying, selling, and trading had been a confusing proposition since
the beginning of time. The fair exchange of cloth, food, or seeds depends
on true standards of measurement. Many standards have been used
throughout history. One of the first was the cubit—the length of a forearm.
Granted, not everyone's forearm was the same, but at least it was always
available. Other body parts served the same purpose. The fathom was the
distance from fingertip to fingertip of a person's extended arms. The span
was the size as a person's splayed hand. Many societies used the foot as
a measure. The Romans divided the foot into 12 *uncaie*, which came be
translated as the words "ounce" or "inch."

The Romans' *mille passus* was a "thousand paces" of its legions as they
marched throughout the world and would, in time, be referred to simply
as a mile. The British used the furlong, originally defined as how long a
furrow could be made by a team of oxen before they needed to rest. Over
time, the furlong became standardized to 220 yards and the Roman mile
to 5,000 feet. During the reign of Queen Elizabeth I, to have exactly eight
furlongs in a mile, the mile was reset to be 5,280 feet.

Despite the wide variety of measurement devices used throughout the
world, it is difficult to imagine any society having a more disjointed system
than France's. Common measures were the *aune, pouce, pied,* and *toise.*

Their actual size would vary by where they were being used. It has been estimated that France had some 800 different measures with total variations numbering perhaps a quarter of a million.[2] The measure of a piece of cloth could depend on the type of cloth. The measure of a parcel of land might depend on the quality of the soil. There were thick manuals created simply to give guidance to those comparing standards in various parts of the country. This would be challenging if everyone were dealing honestly. Some were not. It was a common practice for merchants to buy by one standard and sell by another.

A quest began to find standard measures, beginning with a standard for length—one that was unchanging from region to region and unchanging over time. Any standard would do as long as it held to those two conditions. It could be based on the human foot, but whose foot? It could be what was called the *pied du roi*, the king's foot, though anything having to do with the king was rapidly losing favor in revolutionary France. Various towns selected a standard length, then had it literally carved in stone into a building's wall so townspeople could have a consistent reference. There was a thought that it would be best if whatever was chosen was tied to the natural world. That way, in the future, if a standard was lost or called into question, it could be duplicated by finding it again in nature. The enlightened savants felt this was ideal since it fit with Rousseau's view that society should restore its connection with the natural world.

There would need to be multiple measures tied to this new standard. A foot might be a fine unit of measurements for finding the dimensions of a small field but would hardly be a good choice for stating the distance from Paris to Moscow. Also, rather than a haphazard system, it would be easiest if there was a common rate of conversion. To have 16 ounces in a pound, 48 teaspoons in a cup, 1760 yards in a mile, and a host of other conversions, are a bit much to commit to memory. There was debate as to what that common multiplier might be. It might now seem clear that the multiplier should be ten, but that was not at all obvious to previous generations. Others had been used in the past. The numbers four or eight were considered as possibilities. A fourth of a portion of cloth could be found by folding it over, then folding it again. Folding it over again produced a cloth one-eighth the area. The Spanish recognized the computational advantage of using the number eight with their pieces of eight coin. A base of 12 would be convenient for dividing a commodity into halves, or thirds, or quarters. The ancient Babylonians used a base of 60, which would lead future generations to count 60 seconds in a minute and 60 minutes in an hour.[3]

Numbers such as 12 or 60 are convenient bases when computing with fractions, because they are divisible by many numbers, and fractions were how non-whole numbers were dealt with for most of history. A base of ten is helpful in working with decimals, but decimals were a somewhat recent concept. Fractions had been used thousands of years before decimals made their appearance. At the time of the French Revolution, decimals had been in common use in Europe for only a couple of hundred years. In his *Elements of Chemistry*, Lavoisier sought to convince fellow scientists that decimals were a better computational choice than fractions. There he said, "With this view I have long projected to have the pound divided into decimal fractions [that is, decimals] and I have of late succeeded through the assistance of Monsieur Fourche, balance maker at Paris, who has executed it for me with great accuracy and judgment. I recommend to all who carry on experiments to procure similar divisions of the pound, which they will find both easy and simple in its application, with a very small knowledge of decimal fractions." Ten became the number chosen for this new system, but the person central to the choice of the number ten was an American, Thomas Jefferson.

Jefferson had served as Minister to France in the years 1784 to 1789, but he would continue his admiration for the French when he returned to America. One of the issues that led to the advent of political parties in the United States was a person's view of England or France. Washington, John Adams, and Alexander Hamilton were aligned with England, while Jefferson, James Madison and James Monroe were Francophiles. Jefferson, though American, possessed the spirit of a French savant, as much the creator, inventor, and *philosophe* as any of those active in France at the time. He had his own 28-volume set of Diderot's *Encyclopédie*. He loved all things French—French fashion, French art, French architecture, and French cuisine.

Measurement in North America was scarcely easier than the situation in France. Measuring money was especially difficult. Each colony printed their own currency. The United States' first constitution, the Articles of Confederation, had done little to clear things up. Jefferson, acting as the country's first Secretary of State, proposed a simpler, country-wide system, based on multiples of ten. The dollar would be the basic unit of commerce. There would be ten cents in a disme, 20 cents in a double-disme, ten dismes in a dollar, and ten dollars in an eagle.[4] His basic plan was enacted by Congress and, with a few changes over the years, would become the basis of the U.S. monetary system.

Jefferson also wrote up a plan for measurement that could be used in the United States. Its basic unit was to be the foot and again make use of the number ten, with ten lines in an inch, ten inches in a foot, and ten feet in a decad.[5] However, he held off asking Congress to enact it into law. Jefferson knew that the French were working on a measurement system, and he wanted to see how that turned out. Thomas Jefferson's standard of the foot had its basis in a pendulum swinging back and forth. A swinging pendulum was of interest to the French savants as well.

In the sixteenth century, Galileo Galilei discovered that the period of a pendulum, the time it took to swing back and forth, remained constant. If a pendulum was lifted and let go, it would return in a certain amount of time. Galileo found that if he lifted it from a higher starting point, it still returned in the same amount of time. The pendulum had to travel farther, but it was also traveling faster, and it turned out the greater distance and greater speed balanced. Changing the length of the pendulum changed its period, but that seemed to be the only factor. Neither what it was made of, nor the weight of the bob at the end, changed its period. Dutch scientist Christiaan Huygens made use of this fact to build a time keeping device. Since the only variable was the length of the pendulum, Huygens determined the length that produced a swing of the pendulum equal to one second. The device only needed occasional winding to keep the motion going.

The French and Jefferson both felt this particular length of pendulum, called the seconds pendulum, could be used to create a standard. Because this seconds pendulum had a fixed length, it was proposed that it become the natural standard for measurement. It was suggested it be called a meter, after the Greek word *metron* for "measure." It also happened to be approximately the length of the *aune*, a measure the people were already familiar with. It seemed ideal, and many were in support of it. However, this concept of using a pendulum to establish a length standard would not last. It was found that the period of a pendulum could subtly change depending on the force of gravity. The farther an object is away from the center of the earth, the less the force of gravity acts on it. A pendulum atop a mountain kept slightly different time from one at sea level. Also, because the earth is not a perfect sphere, latitude made a difference as well. Even the mass of a nearby mountain would make a slight difference in a pendulum's timing. Gravity was not the only issue. The temperature and humidity of the air could have an impact. There were enough issues that it began to lose favor with some.

In 1773, Condorcet presented a measurement plan based on the seconds pendulum. It failed to gain acceptance. In 1790, he was a member of the National Assembly, and with the current climate of change, it was a good time to try again. Foreign minister Talleyrand made the formal presentation to the Assembly.[6]

Talleyrand

Charles Maurice de Talleyrand-Périgord was born in Paris in 1754. He had distant parenting, literally and figuratively, spending much of his childhood years away from home. He had a deformed right foot which caused him to limp throughout his life. It was this deformity that caused his father to take the inheritance he naturally would have had as eldest son and give it to his younger brother. He attended seminary and was later named the bishop of Autun. His life, however, did not always reflect the teachings of the church.

He was one of only a handful of bishops that took the required oath of loyalty to the state. He proposed the confiscation and selling of church property to the state to pay off the public debt. For these and other moral lapses, he was excommunicated by the pope. Though not uncommon at that time and place and for those in similar positions, his affairs were more numerous and more blatant than others. He became wealthy through the taking of government bribes. When President John Adams sent negotiators to France, it was made clear no meeting with minister Talleyrand would take place without first making payments to French agents, known as X, Y, and Z. The so-called XYZ Affair led to what was referred to as the Quasi-War, and very nearly open war between France and the United States.[7] Regarding his moral qualities, Honoré Mirabeau, an early leader of the revolution, stated that Talleyrand, "…would sell his soul for money; and he would be right, for he would be exchanging dung for gold."[8]

While he had moral shortcomings, he was an expert negotiator. Not noted as a skilled public speaker, he was very personable and was very persuasive with individuals and in small groups. As Napoléon Bonaparte conquered his way across Europe, Talleyrand composed the treaties with those defeated countries. The Treaty of Vienna was negotiated after the fall of Napoléon's regime, and Talleyrand was an important part of those talks, representing the defeated country of France. Though he had nothing to bargain with, his skill allowed France, for the most part, to simply return to its pre-revolutionary life. In comparison, Germany lost vast amounts of its land and autonomy after each of its defeats in the world wars of the twentieth century.

Regardless of whatever circumstances surrounded him, Talleyrand seemed to come out unscathed. He had the ability to ingratiate himself with the various powers of France that came along. He was able to side with, and hold important positions with, first the monarchy, then the republic, then Napoléon Bonaparte, later the coalition forces that defeated Napoléon, and finally, the restored monarchy.

As soon as Napoléon began making a name for himself as a successful general, Talleyrand began a flattering correspondence with him. Napoléon always recognized Talleyrand's value but would often question his loyalty. In fact, he would likely have agreed that the latter's loyalty was not to Napoléon Bonaparte, but to France. He sought peace and recognized that Napoléon's path of constant warfare would ultimately lead to France's downfall. After the Treaty of Tislit was negotiated with Russia, and before Napoléon's subsequent invasion, Talleyrand met with Czar Alexander in order to frustrate Napoléon's secret plans. As he once told the czar, "Sire it is in our power to save Europe, and you will only do so by refusing to give way to Napoléon. The French people are civilized, their emperor is not."[9]

A Commission of Geniuses

Talleyrand, with technical assistance from Marquis de Condorcet, made his measurement proposal to the National Assembly. The Assembly, along with Louis, still nominally the head of the government, agreed to the plan. Talleyrand's proposal included a provision that the standard of length should be derived from the natural world, the seconds pendulum, and be linked to other standards. This linking of the length standard to that of mass and volume was a fundamental idea of this new system. While accepting Talleyrand's proposal, the Assembly added the requirement that divisions be based on the decimal scale.[10]

The National Assembly established a commission of some of the finest minds in mathematics of that or any age—Laplace, Borda, Condorcet, Legendre, Monge, Lagrange, as well as the chemist Antoine Lavoisier.[11] France had many outstanding mathematicians in its history, including René Descartes, Blaise Pascal, and Pierre de Fermat, but this current collection, all serving on one committee, was unique.

Pierre-Simon Laplace has been referred to as the French Newton. He had taught at the École Militaire and, in fact, been the examiner of one of its prominent students, Napoléon Bonaparte. Laplace had declined Napoléon's request that he accompany him on the Egyptian expedition. Despite the rejection, Napoléon later appointed him to be Minister of the Interior. This, however, lasted just a few weeks when Napoléon removed him from that post and placed him in the Senate. Napoléon either felt Laplace was not up to the task, or perhaps he was just used to hold the spot for Napoléon's brother who replaced him. Laplace was a genius, despite what Napoléon's view of him might have been. He did groundbreaking work in celestial mechanics and probability theory. He used Newton's theory of gravitation to explain the orbits of comets and planets. His book on celestial mechanics would be the standard astronomy text for a hundred years. He worked with Lavoisier on heat theory, and with Italian physicist Alessandro Volta on electricity. Laplace and Volta would die on the same day—March 5, 1827.

Laplace and Joseph-Louis Lagrange were the leading mathematicians in France. Lagrange was the only one of this group not born in France. He was born in Torino (Turin), Italy in 1736 with the very Italian name Giuseppe Luigi Lagrangia. He moved to France after the death of his wife, and at the invitation of Louis XVI. During the Reign of Terror, a law was passed which called for the arrest of anyone in France who had been born in an enemy country. Aware that an arrest at that time often ended in the individual's execution, Antoine Lavoisier intervened and was able to obtain an exception for Lagrange. Laplace and Lagrange largely kept out of the political scene and, for the most part, were able to avoid the turmoil that came to so many.

Adrien Legendre was born in 1752 into a wealthy family, although he lost his wealth during the darkest days of the French Revolution. He was part of a team that conducted a survey between the national observatories in Paris and Greenwich. Legendre did work in the areas of number theory, geometry, and celestial mechanics. He developed an important mathematical concept known as the method of least squares. Experimental data often does not follow an exact pattern. The method of least squares finds a best fitting line or curve that approximates the collected data. This method, still used today, was used by those attempting to determine the earth's curvature.

The mathematician known as Marquis de Condorcet was perhaps the most important individual to the establishment of the metric system, beginning with his initial proposal in 1773, then doing much of the work on its later development. Condorcet was a prime example of one with an enlightenment philosophy. He was a strong proponent of free public education, universal suffrage, and equal rights for women, Jews, and blacks. He was solidly behind the revolution, though not the excesses that came later. He voted against Louis at his trial, but against his execution. Condorcet published political pamphlets espousing his views, but they did not find favor with those in power at the time. A warrant for his arrest was issued, and Condorcet attempted to go into hiding, but he was eventually identified and jailed. The morning after his arrest, he was found lying dead in his cell. It was commonly assumed at the time that he took his own life.[12]

Measuring the Earth

The commission came back to the National Assembly with its recommendation, and it was accepted on March 30, 1791.[13] Its plan had made the switch from using the seconds pendulum. Rather than using the pendulum, the commission decided that the earth itself should be used to establish a standard. What was more unchangeable than the planet earth?

It was proposed that the distance from the North Pole to the equator be found and then one ten-millionth of that amount become the length of the new standard. The one ten-millionth number was chosen because that length would again approximate the already familiar *aune*. This standard was tied to the natural world, so it could be remeasured if there was every any question about its accuracy. Finally, the standard was assumed to be unchangeable. It checked all the boxes. An obvious drawback was the enormous undertaking of measuring that distance. (It would be another hundred years until any human would even get to the North Pole.) However, if a portion of a meridian were accurately measured, the full distance could be easily computed. For example, if the length of an arc was found corresponding to ten degrees of latitude, multiplying by nine would give the pole to equator distance. Even a partial measurement, though, would be a huge project.

Previous attempts had been made to measure France, beginning with the work of Giovanni Cassini. Cassini was born in Italy but was recruited to establish a national observatory in Paris. He was already a famed astronomer, studying Saturn and its rings, as well as discovering four of its moons. (The recent Cassini space probe sent to explore Saturn was named for him.) In the seventeenth century, he conducted a major survey of France—the first time in history that a country had attempted such a survey of itself. It was used to establish the country's boundaries more accurately and found that France wasn't quite as large as previously thought, leading Louis XIV to complain to him, "Your journey has cost me a major portion of my realm."[14] One result of the survey was that it led Cassini to believe that the earth's circumference was longer through the poles than through the equator.

The earth was not flat. That had been known for centuries. It was assumed, however, that aside from its mountains and valleys, it was perfectly spherical. Isaac Newton, applying his theory of gravitation, believed it was not. Because of its rotation, he thought the earth had an equatorial bulge. French savant René Descartes believed the opposite. He theorized that the greatest distance through the earth was from pole to pole. Who was correct would be a topic of debate, once again falling largely along nationalistic lines. If this new meter would be based on the size of the earth, this was a crucial matter.

France sent out two more expeditions to attempt to make a final determination. One was sent to Peru, measuring near the equator, and one was sent to Lapland to measure near the Arctic. If Newton was correct, a degree of latitude would be longer at the equator than it would be near the North Pole. The method used in these early surveys was much the same as what would be employed when establishing the meter. First, a baseline was established by selecting two landmarks to be its endpoints. Star sightings were used to find the latitude and longitude of those points. To obtain good results, the baseline needed to be several miles long. Several 20-foot-long wooden poles were repeatedly laid end to end to find its length.

By using the endpoints and then selecting a third landmark, a triangle was formed. A quadrant, or similar device, was then used to find the angles adjacent to the baseline. These quadrants, devices so named because they were in the shape of a quarter of a circle, were made of heavy iron and were approximately three feet across. Angle measures could be found using other devices such as sextants, a sixth of a circle; octants, an eighth of a circle; and later, a device known as a repeating circle. With two angles and their included side known, trigonometry could then be used to find the lengths of the triangle's remaining two sides. Once all the information was obtained for that triangle, another landmark point was chosen. By selecting one of the sides found on the first triangle and this new point, a second triangle was constructed, and the process repeated, using a quadrant to find the angle measures, and then using trigonometry to find the lengths of the sides. Using this process of triangulation, a string of triangles could be constructed indefinitely.

The results of the equatorial and Arctic surveys showed that Newton was indeed correct.[15] Due to the earth's rotation, the earth bulged along the equator. This was just one more complication that would need to be accounted for in the survey to establish the meter.

The commission had determined conditions for where this survey should take place. It must cover at least ten degrees of latitude, and to minimize the earth's eccentricity, straddle the earth's line of 45 degrees latitude. The endpoints must be at sea level. Also, this survey must cover a route that had already been at least partially surveyed.[16] It turned out that France was the spot that best fit these conditions. The fact that the conditions selected by the French committee just happened to be satisfied only by France seemed just a bit too convenient to other countries. This, along with France's decision to drop the seconds pendulum standard, caused several countries to lose interest in the French plan. That the French were attempting to keep the spotlight on themselves is perhaps demonstrated best by Talleyrand's claim that this showed "…that in this field, as in many others, the French Republic is superior to all other nations."[17]

The committee met with the king regarding the new survey. One of the committee members, Jean-Dominique Cassini, was the great grandson of Giovanni Cassini. Giovanni's descendants had all followed in his astronomical footsteps, taking turns at directing the national observatory. The king asked why another survey was even necessary since it had already been accomplished by his great-grandfather. The great-grandson told him the instrument to be used would be much more accurate than what was used in the past. Louis gave his blessing to the project. Cassini would later be imprisoned for what others felt were ties too close to the monarch.[18]

This more exact measurement device was known as the Borda Repeating Circle. It was named for Jean-Charles Borda, one of France's leading physicists and mathematicians. He had also been a naval commander and oversaw several French ships sent to help the American colonies gain their independence. He was captured by the British, though later released. Borda dabbled in many different areas, including the development of a new method of tabulating votes cast during elections. His repeating circle weighed approximately 20 pounds and could make horizontal or vertical readings. Previous instruments were accurate to within approximately 15 seconds of angle measure. The repeating circle was given this name because after a measurement was taken, the device was rotated and remeasured, each time increasing its accuracy, making it accurate to within approximately one second.[19] The survey team took repeating circles with them that were calibrated to both 360 and 400 degrees. The 400-degree circle was another innovation of the French during this time. The 400-degree circle was not a bad idea. In the new system, right angles would be 100 degrees rather than 90, and every degree of latitude would be exactly 100 kilometers.[20] However, it was an idea that did not catch on.

Delambre and Méchain

In 1792, Jean-Baptiste Delambre and Pierre Méchain launched the survey which would determine the length of a meter, with Delambre starting in the north and Méchain in the south. Jean-Dominique Cassini was to join them but opted out due to the recent death of his wife and faced with having five children in his care. Adrien Legendre had also been chosen to head a team but dropped out so Delambre could have his spot.

Delambre and Méchain were roughly the same age, were both educated by Jesuits, and had both studied under Jérôme Lalande at Collège Royal, now known as the Collège de France. Lalande was a believer in enlightenment thought, which included being an atheist. His atheistic writings would be troubling to Napoléon when the soon-to-be emperor made a show of bringing the pope to his coronation. Despite his unbelief, he had taken in several priests that faced possible execution, under the guise of them being fellow astronomers. He also provided to du Pont de Nemours a place of sanctuary from those who sought him. Lalande was an accomplished astronomer, cataloguing over 40,000 stars, computing the orbit of Halley's comet, and calculating distances to the moon and the sun. Lalande taught at Collège Royal for almost fifty years, and when he passed away, years later, Delambre filled his position.

Delambre was nearly blinded by smallpox as a child. Knowing he might face total blindness motivated him to read and study all he could. His vision improved as he grew older, but it would always be a problem for someone whose life's work was based on observation. Many scientists of that time had patrons that helped finance their work. His was Geoffrey d'Assy. Delambre had tutored d'Assy's son and would become part of the family, living with them many years. D'Assy supported Delambre's work financially and even built him his own observatory. D'Assy was later arrested for royalist leanings. After being imprisoned for five months, he was executed with fifty others who had been accused of planning a prison riot.[21]

Méchain had experience in both surveying and astronomy, was part of a mission to map the French coast and was the discoverer of 11 comets. Like Delambre, he also found himself in need of financial help. His wife's family had previously been employed by the royal court, but with the coming of the revolution, what money they had was lost. While Méchain was away on the expedition, his wife, Barbe-Thérèse, continued his astronomical observations in order to maintain an income for the family. She did this for the duration of his journey, one that was supposed to last months, but instead, lasted years. It was not always an easy task. Two days after the storming of the Bastille, 300 revolutionaries invaded their observatory looking for gunpowder. They found none, but the event understandably left her shaken.

The two survey parties set out from the opposite ends of the country and planned to meet in the middle, at the town of Rodez. The northern portion was longer, but also less mountainous. Like the previous surveys, a baseline would be identified along with landmarks defining the triangles. Angle measurements would be made, then lengths of the triangles computed. There would be complications. The vertices of the triangles were at various altitudes and had to be, in essence, leveled mathematically. Additionally, because the earth is a curved surface, the geometry itself was different from the Euclidean, or flat-surface, geometry they would have been most familiar with. In sighting over long distances, the atmospheric refraction of light had to be considered. Sites previously used in the original Cassini survey were no longer there or no longer visible. For one of the baselines, 600 trees had to be pruned.[22] Stars were used to determine locations, but the weather didn't always cooperate. Delambre noted that, in one two-month span of time, only two occasions were suitable for making nighttime observations.[23] At one point, a pack of wild dogs invaded and trashed their camp.

Those are just issues that the natural world caused. People brought their own set of problems to the project. Spying, either by countries at war with France or by elements within the French government, was not uncommon. Early on, Delambre and his crew were taken into custody on suspicion of being spies. They were able to produce papers showing their work was indeed approved by the government. The fact that that approval was given by a French government no longer in power complicated the issue, but it was eventually resolved and they were released.

At night, the location of landmarks could be viewed when a fire was lit, backed by a parabolic reflector marking the site. There was confusion at one point when the Tuileries Palace, housing the king at the time, had been set on fire by a riotous mob the same night they were attempting an observation.

There were additional issues specific to the southern part of the survey. The meridian line they sought to measure met the sea just across the border in Barcelona, Spain. Initially, Spain welcomed a survey crew because the exact location of the French-Spanish border had been unclear, and this project could remedy that. However, tensions between the two countries were increasing to the point that Méchain's team was ordered to abandon the location where they were working and told to dismantle the observatory that was set up there. Their funds were impounded by the Spanish government. The following month, Méchain was severely injured when inspecting a pumping station. In a freak accident, a beam had become dislodged, striking him in the chest. He was fevered and unconscious for three days and bedridden for a period well after that.[24] Spain eventually allowed Méchain to leave the country to recuperate in Genoa with his promise he would not share with France any information he had collected.

Meanwhile, in northern France, Delambre's work also came to an abrupt halt. On August 8, 1793, the Academy of Sciences and other academic societies in France were shut down. The only learned organization allowed to continue was the Commission of Weights and Measures, though even they lost key members. The reason? A letter from the Committee of Public Safety explained, "...considering how important it is for the improvement of public morale that government officials delegate their powers and functions solely to men known to be trustworthy for their republican virtues and the abhorrence of kings ... [The committee] decrees that from this day forth, Borda, Lavoisier, Laplace, Coulomb, Brisson, and Delambre cease to be members of the Commission of Weights and Measures, and that they immediately hand over to the remaining commissioners all their instruments, calculations, notes...." Those that were left were to, "apply Revolutionary enthusiasm to bring the new weights and

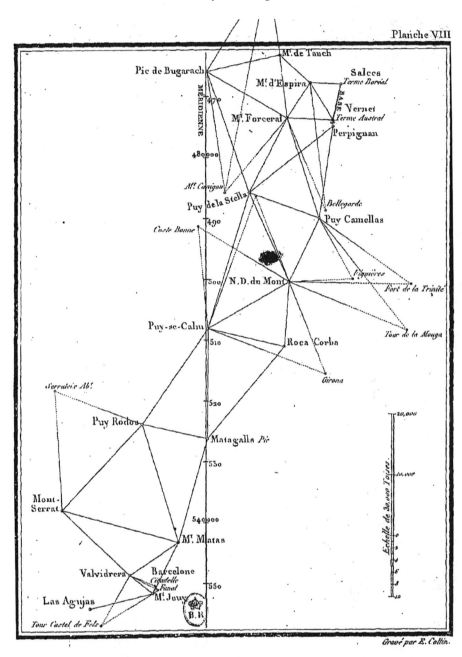

The diagram shows a portion of the triangulation work done by Delambre and Méchain as they surveyed the length of France. Their work would lead to the establishment of the length of the meter.

measures into use among all citizens."[25] Joseph Lagrange, who managed to steer clear of factions during the revolution, was appointed the new chairperson of the commission.

Eventually, both Delambre and Méchain were able to resume their work. Méchain, however, would never be the same, either physically or emotionally. One of his readings at the southern end of the meridian was wrong. It did not match the rest of the data he had already collected. He likely could have discovered the source of his error, but due to the increasing tensions between France and Spain, his Spanish hosts had granted him only a single day to complete his work there. It has never been determined what went wrong; whether it was his mistake, a problem with the equipment, or some other issue, but Méchain knew his data was wrong and it would haunt him the rest of his life.

His depression caused him to make little or no progress after he resumed his work. Delambre wrote to him repeatedly, asking him how things were going and if Méchain would send him the data he had collected so far. When Méchain did respond, which was rare, it was only in vague terms. At one point, he showed his state of mind, stating, "I have spent all my time in the cruelest anxiety, unable to concentrate on what I am doing, continually reproaching myself for the past because the present is unbearable and because I tremble for the future."[26] Altogether, Méchain lost over two years due to the forced stoppage of his work by the Spanish government and time spent recuperating from his injury. From then on, his progress was slow or, at best, sporadic. His thoughts often turned dark. "After all that has happened, I can no longer show myself anywhere and my only wish is to be annihilated."[27]

His long-suffering wife, Barbe-Thérèse, was aware things were not right with her husband. She had been exchanging letters with him and Delambre. Though she had taken over much of his work at home and was caring for their children, she now resolved to travel to where he was to meet him face to face. Their meeting would be the first time she had seen him in six years. Barbe-Thérèse left her family in the care of others and made the journey to the south of France.

Following their time together, she wrote Delambre, "Citizen, having completely failed in my mission, and with a heart pierced by a thousand griefs, I return to Paris.... I have begged him in vain to write to you and to agree to measure the baseline of Perpignan with you. Not wishing to upset me, he always answered vaguely. For the first time, my husband has dissimulated with me.... I urged him not to throw away the fruits of all his years of suffering and sacrifice for something so idiotic. I decided to force the issue. I insisted that I would not leave his side until he had finished all his triangles and had been reunited with you. He thereupon renounced the baseline irrevocably, and vowed that he would concede all the glory to those whom fortune had favored.... I wrote to Borda and explained my situation, no longer able to withstand the horrific blows which assail my heart. I know my husband to be a man of talent and virtue.... It is the extreme sensitivity of his soul which has ruined him. He is more unhappy than blamable. I have obtained from Citizen Méchain a promise that he will not break off his labor until his triangles are complete and the results sent to you and Borda. It's all I could do. My family calls me back to Paris, and I cannot stay here until the mission is done. Madame Méchain."[28]

Though Méchain was still of troubled mind and working far too slowly, his wife was able, with much prodding, to increase his efforts. Ultimately, Lalande told Delambre to go south and help him finish the work. Delambre did so and eventually the work was completed. With the work finally done, Delambre was assigned the job of writing a final report. To deal with his contradictory results, Méchain ultimately chose to simply make up numbers that fit with the rest of the data. On close examination, Delambre was able to see the attempted subterfuge. Rather than expose him, Delambre wrote his report and stated that the underlying data would be stored in the archives of the observatory, knowing one day his report would be examined. It was, though not for a hundred years. At the time, he wrote, "I deposit these notes here to justify my choice of which version of Méchain's data to publish. Because I have not told the public what it does not need to know. I have suppressed all those details which might diminish its confidence in such an important mission, one which we will not have a chance to verify. I have carefully silenced anything which might alter in the least the good reputation which Monsieur Méchain rightly enjoyed for the care he put into all his observations and calculations."[29]

Until that time in the distant future, they were both considered national heroes. Méchain was made director of the national observatory. By then, his oldest son was already in his father's scientific footsteps, following General Bonaparte to Egypt. Méchain, perhaps to assuage his guilt, sought to revisit the survey years later. Rather than placing the terminus in Spain, he placed his baseline in the Balearic Islands, south of Barcelona, in the Mediterranean. There, he fell ill with malaria and died on September 20, 1804.

Delambre was appointed to prestigious positions and received many honors. At the fall of Napoléon, he lost some of the honors given him, but they were restored by Louis XVIII. On August 19, 1822, Jean-Baptiste Delambre passed away, peacefully and honored by his country.

Perhaps the greatest irony is that the basic assumption underlying the expedition was wrong. It was assumed the distance from pole to equator passing through France was the same as the pole to equator distance going through any other region. However, it was found that the earth is not symmetric. Scientists later discovered that, in addition to the bulge along the equator and various mountains and valleys, there are other random warpings of the earth's surface. The warpings are not extreme, but enough that the resulting distances were not the unique number the savants assumed it would be.[30]

Regardless of whether it was unique or not, there was now a number for the distance from pole to equator, and one ten-millionth of that length would be called a meter. A law was passed on March 30, 1791, officially accepting it as France's standard length. In succeeding years, other legislation would refine this new metric system. The prefixes, such as milli-, centi-, deci-, all based on powers of ten, would be added. This length standard was linked to the concepts of area, volume, and mass. This was a new concept. For example, in the English system, a foot has no connection to pints and pounds. A liter was to be a measure of volume, equivalent to a cube with each side being one tenth of a meter, or in other words, a cubic decimeter. The unit of mass, called a gram, was developed by Antoine Lavoisier and would be that of one cubic centimeter of water at its maximum density. Substances other than water have their maximum density in their solid state. Water's greatest density was found by Lavoisier to be at positive four degrees Celsius and was what would be used to establish the standard for mass.

A decision was made that prototypes of the meter and the kilogram should be constructed. Prior to the revolution, Marc Étienne Janety was employed in the royal court as a goldsmith but moved to Marseilles in the south of France when all things associated with royalty became endangered. Those times had now passed, and he was called back to construct a platinum cylinder of equal height and diameter and with a mass of exactly one kilogram. The one-kilogram cylinder has only recently been superseded, with a definition now based on Planck's constant. The meter has had a bumpier journey. A platinum bar was constructed by Alexander Lenoir in 1799, which, in 1889, was replaced by one made of a platinum-iridium alloy. In 1960, the meter was redefined to be a wavelength of a spectral line of krypton-86. Since 1983, the meter has been defined as the distance light travels in 1/1,299,792,458 of a second.[31] The original artifacts still exist but are only historical relics now.

A Worldwide System

In 1798, Talleyrand invited countries to what could be called the world's first international scientific conference.[32] Its goal was the adoption of the metric system, not just in France, but in other countries as well. Despite all the efforts put into this endeavor, in 1812, Napoléon decided to scrap the whole thing.[33] He, as well as much of France, never really learned the system. Also, it was not totally embraced by other countries, causing Napoléon to feel it was interfering with trade.

The metric system, however, would make a comeback. Ironically, the first to fully adopt the metric system were countries other than France. By 1820, the nearby countries of Belgium, Luxembourg, and the Netherlands had officially taken it as their system of measurement. In 1837, France joined them, restoring what it had created then discarded. It was made mandatory throughout France in 1840.

From the beginning, French savants had hoped for the universal adoption of its new method of measurement. Despite a rough beginning, there has been a steady parade of countries adopting the system since the middle of the nineteenth century—Spain in 1858, Italy in 1863, Germany in 1872, Russia in 1925, and India in 1954. Thirteen African countries adopted it in 1960. Britain, a French enemy for centuries but now on much friendlier terms, joined in 1965. Now, almost 200 countries use the metric system. Currently, the United States is one of three countries not to make the change. This is in spite of the difficulties with what is called the imperial system and the confusion of trying to live in both systems. Doing so entails having to deal with gallons and imperial gallons; miles and nautical miles; and long, short, and metric tons. Ounces are completely different concepts if measuring weight (16 ounces in a pound) or volume (16 ounces in a pint). It was this confusion that was responsible for the loss of the 125-million-dollar Mars Climate Orbiter, burned up in Mars' atmosphere, because the orbiter was measuring force in newtons and Houston Mission Control was measuring force in pounds.[34]

But the United States has come close to its adoption. In 1794, Joseph Dombey, a French mathematician, was sent to the United States with models of the meter and the kilogram to attempt to educate and gain support. A storm caused the ship he was on to land on the Caribbean Island of Guadeloupe, far removed from its target. Guadeloupe was a French colony, and like France, had its share of turmoil among royalist and anti-royalist factions. Dombey was imprisoned, though eventually released. On his way again, his ship was taken by British privateers. Dombey died in a British prison on the island of Montserrat.[35]

Within the United States government, there have been proponents of changing entirely to the metric system, and much potential legislation has come and gone, but seemingly never closer than with the passing of the 1975 Metric Conversion Act. However, the act was not binding, with no enforcement provision. The act was, in effect, ignored by the following administrations.

The French sought to also change how time was measured, though their sweeping changes ended up being very unpopular. The French people were told that there would now be ten hours in a day, a hundred minutes in an hour, and a hundred seconds in a minute. (The new republican second was shortened to be a fraction of the traditional second.) This, of course, meant changing all the timepieces, which for the most part, was simply not done. Of all the measurement changes made, this was the least popular and the first to go.

The French also set out to change days, weeks, months, and years. A new calendar was primarily the brainchild of Gilbert Romme. He developed the calendar in 1793, but within two years, he would be dead. He was active in the revolutionary political scene and became a member of the Legislative Assembly. Making the same mistake as so many others, he aligned himself with the wrong group at the wrong time and was arrested. He was convicted and sentenced to face the guillotine for being a Jacobin.[36]

His calendar plan was implemented, though ultimately did not last. First, the concept of BC and AD with their religious connotations (Before Christ and Anno Domini—in the Year of the Lord) had to go. Time would be measured from the date September 22, 1792, which was the beginning of the new French Republic.[37] This was additionally pleasing as it tied the calendar to the natural world by coincidentally falling on the date of the autumnal equinox, a time of year when day and night are exactly the same length. It almost seemed to be an act of God, if they believed in such a thing. There would still be 12 months in a year, but each contained exactly 30 days which was made up of three ten-day weeks, referred to as *décades*. That totaled to 360 days, so there were an extra five days (or six during leap years) added on, declared to be festival days of patriotic celebration.

The names of the months were changed. The months being named after dictators such as July (Julius Caesar) or August (Augustus Caesar) or gods, even if they were clearly mythological (Janus, Mars, Maiesta, Juno), would be discarded. New names would possess a suffix identifying its season: -aire for fall, -ôse for winter, -al for spring, and -dor for summer. The first day of the new Republic, September 22, 1792, retroactively, became 1 Vendémiaire in the Year 1. The other names for months would celebrate nature. The months of the year were Vendémiaire, Brumaire, Frimaire, Nivôse, Pluviôse, Ventôse, Germinal, Floréal, Prairial, Messidor, Thermidor, and Fructidor. Their names were ones describing its time in the year, such as Floréal being the month of flowering. Some of the others referred to months of frost, snow, heat, or harvesting.

Those names for months are no longer used, but still exist as the names of events during the French Revolution. The Coup of 18 Brumaire (the 18th day in the month of Brumaire) brought Napoléon Bonaparte to power, the Law of 22 Prairial was enacted by the Committee of Public Safety at the height of The Terror, and the Thermidorian Reaction was the counterrevolution following the fall of the Committee of Public Safety. The days of the week are often tied to ancient religious mythology (Tiew's day, Thor's day). They received new names as well. The new names for the days of the week; e.g., Primidi (first day), Duodi (second day), Tridi (third day); not only shed any religious significance, but hopefully a ten-day week would also make keeping the Sabbath more difficult.

It turned out that there were more things to measure. Some changes were successful, and some were not. The Treaty of the Meter was signed in Paris in 1875 by 17 countries, ironically including the United States. It would lead to the creation of the SI, or Système International, which would come to encompass seven different measures with standard units assigned to each: length (meter), mass (kilogram), time (second), electric current (ampere), temperature (degrees Kelvin), amount of substance (mole), and luminosity (candela).

The basic unit of electricity was named for André-Marie Ampère, born in 1775 in Lyon, France. His father, being a follower of Rousseau's philosophy on education, had André basically teach himself. His father was not in favor of the actions of the Jacobins, resulting in his being sent to the guillotine. In his sorrow, André, 17 years old at the time, simply ended his studies for two years. He would later go on to a successful teaching career and do important work, demonstrating the relationship between electricity and magnetism.

Though not a measure in SI, money is also a type of measure. During the French Revolution, the old units of currency were discarded, and the French franc was adopted. Again, it would be based on the number ten, with ten *centimes* in a *décime*, and ten *décimes* in a franc. The franc was copied by many of the countries that were taken over during the Napoléonic Wars, including Switzerland, Belgium, Luxembourg, and several countries in Africa.

Not every aspect of the metric system lasted, but what did, changed the world. As Napoléon once said regarding this system, "Conquests will come and go, but this work will endure."[38] It has so far, for all except Liberia, the United States, and Myanmar.

V

Napoléon Bonaparte

Napoléone di Buonaparte was born August 15, 1769, on the island of Corsica. The island is just off the southern coast of France, only thirty miles from the island of Elba, where he would be exiled 45 years later. Corsica had been under the control of Genoa but was sold to France the year before Buonaparte was born. His parents, Carlo, a lawyer, and Letizia Buonaparte, had eight children who survived to adulthood. When their famous brother came to power, several of his siblings and various in-laws would be given important positions, even heads of countries, though with mixed results.

The Rise to Power

Most of Buonaparte's childhood was spent on Corsica, but because of his family's status, he was able to go to France to attend school. The adjustment was sometimes difficult for him, and there are reports of his being the brunt of jokes. He sounded different because of his accent, though he would try to fit in better by adopting the more French appearing name "Napoléon Bonaparte." Possibly, he was also teased because of his stature. Many have claimed he was of average height for his time. However, after he achieved fame, he was commonly called The Little Corporal. He was also referred to by others (sometimes as a term of endearment and sometimes not) as "our little cropped one,"[1] "Little Boney,"[2] "the little man,"[3] and "that little bugger."[4] Letters referred to "his small size"[5] and being "very little."[6] One of his generals, General Kéber, upset when Napoléon deserted him in Egypt, called him "that Corsican runt."[7]

His education in France included school in Autun, and military academies at Brienne and the prestigious École Militaire in Paris. He was a good student and especially strong in math. He once said, "To be a good general you must know mathematics."[8] He was an avid reader and very familiar with the enlightenment authors. Bonaparte trained to be an artillery officer and completed his studies a year ahead of schedule. His military career, however, was interrupted with trips back to Corsica where he involved himself in the politics of the island, as well as caring for his family after Carlo passed away.

Bonaparte's first military success came at Toulon on the French Mediterranean coast. The First Coalition had formed, and the British had taken possession of this important port city. French troops had made attempts to retake the port but without success. Bonaparte devised a plan to take key sites on the hills overlooking Toulon and place artillery on those locations. In the ensuing battles, he fought alongside his troops, receiving a bayonet wound as he did.[9] After taking those positions and installing his artillery units, he bombarded the British troops below. The British sailed away and the young artillery officer, Napoléon Bonaparte, soon to be General Bonaparte, had his first taste of fame.

Bonaparte was put in charge of the Army of Italy to fight a campaign that, thus far, had not been going well. But first, he would marry Josephine Beauharnais. She had previously been married to a general in the French army. The general, and later Josephine, were arrested. He was arrested because his performance as an officer was deemed unsatisfactory, and she, seemingly for not much more than guilt by association. He went to the guillotine, but Josephine was later released after the fall of Robespierre. She met Bonaparte, by then a rising military officer, and the two were married. It would be a difficult relationship with both being unfaithful, Josephine seemingly more so. They would divorce ten years later, according to Bonaparte, because she failed to produce an heir.

They married on March 9, 1796, and two days after the wedding, Bonaparte left for the front. After victories in Lodi, Arcola, and Rivoli, Bonaparte's troops were soon in control of the region. He followed those military victories by asserting his will on the conquered territories. Though it was not his place to negotiate treaties, create new republics, or write their constitutions, that is what he did, then simply notified Paris for its rubber stamp. He sent to Paris countless Vatican and other Italian treasures that would later be displayed in the Louvre Museum. Thus began what would become his practice of helping himself to the treasures of a conquered land.

He was a hero when he returned to Paris. He was given the charge of putting down an uprising in the city. He did so, having his troops fire into the crowd, killing hundreds of the rioters.[10] While not appreciated by the insurrectionists, it gained him support within the government.

Bonaparte had dreams of going to Egypt, and the French government, not needing someone more popular with the people than themselves, agreed. He would return two years later, seemingly victorious, although not as victorious as he claimed. Seeing his continued popularity, he was soon recruited by French foreign minister Talleyrand and Emmanuel-Joseph Sieyès to orchestrate a coup of the French government. Sieyès, who had been instrumental in establishing the National Assembly in 1789, was also a member of the Committee for Public Safety, the Council of 500, and now the Directory. For his next step, he needed, as he termed it, "a sword" to strengthen his takeover; although in selecting Napoléon Bonaparte, he got more than he anticipated.[11]

The Coup of 18 Brumaire (November 9, 1799) began by convincing or coercing the other members of the five-man Directory to resign. Napoléon, along with a group of his soldiers, then entered the legislative chambers and told the legislators that he must take over for their own protection and the good of France. Because this pronouncement did not sit well with the members, he was heckled, assaulted, and thrown out. More troops were gathered and brought in to restore order and legislators scattered. Napoléon's cause was helped by the fact that his brother Lucien just happened to be the president of the Council of 500. For those members who remained, Lucien led them in voting to dissolve the Directory and the legislative councils. A new constitution would soon replace the five-person Directory with a three-person consulate, with Napoléon established as First Consul.

With his position at the head of government secured, he headed east to restore his previous military gains. Now fighting a second coalition of countries, Napoléon secured a major victory at the Battle of Marengo, culminating in a treaty and a rare period of peace until the next round of fighting.

For the next 15 years, Napoléon would lead the country of France, but also lead its armies to victories at Ulm, Friedland, Jena, Wagram, and probably his greatest, at Austerlitz, a victory he celebrated by constructing the Arc de Triomphe. After signing a peace treaty with the countries of the Second Coalition, Napoléon returned to France and his role as First Consul. According to the new constitution, he would serve a term of ten years. Later, by proposing changes to the constitution which were confirmed by plebiscite, his position was transformed from serving a ten-year term, to that of a First Consul for life, and then emperor. There would still be a legislature, and France would still, confusingly, be referred to as a republic,[12] but in practice, it was Emperor Napoléon holding the reins of government. Votes of the people to approve these constitutional changes were overwhelmingly positive—implausibly so. For one vote, Napoléon received a healthy 99.95 percent approval.[13] While the vote totals are difficult to believe, it is also difficult to argue with success, and he was successful at home and on the battlefield. His elevation to emperor, however, was not popular with everyone. Ludwig van Beethoven had dedicated his Third Symphony, *Eroica* (Italian for "heroic"), to Napoléon, then took back that dedication after Napoléon took on the mantle of emperor.[14] Simón Bolívar, the liberator of South America, was in Paris at the time and said, "He made himself emperor, and from that day on, I looked upon him as a hypocritical tyrant, an insult to liberty and an obstacle to the progress of civilization."[15]

Napoléon the Progressive

Perhaps his greatest and longest lasting accomplishment was the Napoléonic Code. Previously, there was a jumble of 366 different, often contradictory, codes of laws in the various cities and regions of France.[16] Being governed by a different set of laws as one traveled or did business through France was one of the grievances brought to the Estates General years before. This new comprehensive code would cover areas such as inheritance rights, marriage rights, property rights, and grounds for divorce. There were 102 sessions needed to write this new code and Napoléon somehow found the time to preside over 57 of them.[17] In time, this code would not only govern France but would be imported into those areas conquered by France, as well as being adopted by many other countries. This code of laws would survive as the basis of most civil law throughout the modern world.[18]

Napoléon sought to restore relations with the church, though he did so primarily to serve his own political ends. He anticipated this restoration increasing his popularity with the people of France. As he said, "I intend to re-establish religion, not for your sake, but for mine."[19] The Concordat of 1801, an agreement reached between Napoléon and the Catholic Church, reversed a great deal of the rejection of the church that had taken place in revolutionary France. The concordat allowed the church to resume its previous role in French life, though with the government (essentially, Napoléon) still controlling much of how it functioned. It declared Catholicism to be "the religion of the great majority of the French"—quite different from restoring it to the position of being the recognized state religion, but certainly an improvement on its status during the French Revolution. In restoring the seven-day week, it brought back the religious concept of the Sabbath. The Cult of Reason was ended, as was the irreligious and very unpopular revolutionary calendar. The concordat reestablished the relationship between the church and the country and would last until 1905 when new laws established a separation of church and state.

There were gaping holes in education because many of the schools had closed during the revolution when those in power had limited the reach of the church. Several individuals, including the philosopher and mathematician Marquis de Condorcet and the chemist Antoine Lavoisier, submitted proposals for major changes, but in a time of chaos, those proposals were lost in the political confusion. While religious schools would make a return, a more cohesive system of public education was established by Napoléon. Though primary education remained relatively unchanged, the secondary level saw the establishment of the *lycée*. In 1801, Napoléon established *lycées* throughout France to be a more rigorous educational structure, serving as a preparation for the university level. Napoléon made certain that, at the same time, they were beneficial to the state. They provided training for work in the government or the military—having uniforms, marching drills, and appropriate course work that would advance the goals of the nation.

Another painting done by Jacques-Louis David. Napoléon Bonaparte's interest in mathematics and science was the impetus for much of the discovery that took place in France. Napoléon reshaped Europe, opened Egypt to the world, and established a code of laws for France that was adopted by many other countries.

The university level was strengthened, again often with a primary goal of benefiting the state and its military. The École Polytechnique was created by the National Convention in 1794 and was transformed into a military school by Napoléon. The Sorbonne, which had been vacated during the revolutionary era due to its religious emphasis, was restored.

The state of the economy was one of the major causes of the French Revolution but began to improve under Napoléon. There had been certain improvements in the 1790s due to better harvests. Also, the manpower needed for the armies to fight France's wars supplied jobs to the populace. The looting of conquered lands supplemented the treasury as well as adding to the collections of French museums. Napoléon was even able to balance the budget, something that had eluded the French government for years. However, the underlying structure was still troubled. Napoléon established the Banque de France, which is still in existence, stabilized France's currency, and did much to put France on a solid economic footing. Roads, canals, and hospitals were built or improved.

Few people are military as well as political geniuses, but Napoléon's victories are still studied at military schools throughout the world. His use of speed, surprise, and aggressiveness did much to overcome disadvantages he might have had numerically. He was adept at concentrating his own troops at opponents' weak points. He sought to divide his adversary and attack those pieces separately. Warfare of the time was often two complete armies hurling themselves against each other, but Napoléon often maintained a portion of his most effective troops in reserve to be called upon during crucial points of battles. He developed the military corps concept. Rather than an entire army being divided into infantry, artillery, and cavalry; smaller corps units were established containing all three, increasing mobility. Lord Wellington had stated that that he would rather have heard an opposing army was reinforced with 40,000 men than to hear that Napoléon had arrived to take command.[20]

Napoléon's armies changed the map of Europe and the world. The Holy Roman Empire (which has been described as being none of those three concepts) had been in existence for a thousand years. It now ceased to exist. Countries such as the Duchy of Warsaw, the Confederation of the Rhine, the Cisalpine Republic, the Kingdom of Holland, and Westphalia were newly created. French involvement in wars and treaties caused lands in North America to be transferred between countries. Simón Bolívar, the liberator of South America, was educated in France and was present at Notre Dame for Napoléon's coronation.[21] He was greatly influenced by French Enlightenment authors and could view firsthand the effects of France toppling its monarchy.

Not all the actions of Napoléon would be considered progressive. He used the press to further his purposes, simply closing down 60 of the 73 newspapers and controlling the news in those remaining.[22] The devastating defeat in the battle of Trafalgar wasn't even initially reported, and when it was, it was claimed as a French victory.[23] He was skilled in the art of subterfuge. As he said on more than one occasion, "In this world one must appear friendly and make many promises, but keep none."[24] France's role in the slave trade was ended during the years of the revolution, but Napoléon reinstated it in order to harvest the lucrative crops in the French-owned islands of the Caribbean. Women lost rights during the Napoléonic Era. Napoléon once said, "Women should not be looked upon as equals of men. They are, in fact, only machines for making babies."[25]

Throughout his career, Napoléon was shrewd and seemed to make few mistakes in judgment; however, mistakes seemed to come more frequently in his later years. It is possible there were underlying health issues. He had clearly gone from being a young man with a slender build to being a heavy-set, middle-aged one. Perhaps the stress of running an empire was catching up to him. Whatever the cause, events, and his response to them, began to take place that would ultimately lead to his downfall. By his own choosing, he fought a war on two fronts, in Spain and in Russia, trying to enforce an ill-conceived plan, the Continental System, that had little chance of succeeding. He invaded Russia and was drawn further and further into the country—then too late, decided to retreat into the teeth of a Russian winter. When he did retreat, he chose a northern route, when a warmer, southern one might have saved his army. As he fought the collective powers of Europe, he rejected peace overtures that could have left him in power and left France with roughly the same borders it had previously.

The Continental System

England controlled the seas. The Battle of the Nile in 1798, and then the Battle of Trafalgar in 1805, off the coast of Spain, had seen to that. While Napoléon could not defeat England's navy, he did have control over his allies. His Continental System was a plan to shut down the trade of France's allies with Britain. This embargo, which affected most of the continent of Europe, did hurt England, but also caused a great economic burden to those participating countries. Britain sought to maintain its naval superiority by breaking the embargo with its own blockade of French trade, and by boarding other countries' ships—practices that led directly to the War of 1812 with the United States.

There would be leaks in this Continental System, notably in Portugal and Russia. Napoléon sent troops from France, through Spain, to put a halt to Portugal's violation of the embargo. Spain had been an ally of France, but the incursion of troops, and Napoléon's placing of brother Joseph on Spain's throne, caused a serious rift in their alliance. In Spain, a guerrilla war was waged against the French troops. (*Guerra* is Spanish for "war," *guerrilla* is "little war," and in this campaign, the term "guerrilla war" was first used.) Additionally, British troops under Wellesley, soon to be known as Lord Wellington, landed in Portugal and engaged French troops. He would, in time, cross the Pyrenees and enter France itself.

Russia and the End

In 1807, Napoléon drove through Europe with a series of victories. This led to Russia's Czar Alexander meeting with Napoléon on a barge in the middle of the Nieman River to sign the Treaty of Tilsit. It was a treaty that did not simply cease hostilities, but also committed Russia to participate in the Continental System. As time went by, Russia would lapse in the agreement just as Portugal had done, and as with Portugal, Napoléon would invade.

He brought approximately a half-million men, many of them allied troops, into Russia in June of 1812. It was the largest army the world had ever seen.[26] Napoléon's goal was not to capture land, but to defeat Russia's army and enforce his blockade. He was not able to do either. Napoléon drove deeper and deeper into Russia, fighting an enemy that simply kept drawing him in. The massive battle of Borodino, numerically a French victory, saw Russian troops simply retreat and regroup. The French army was able to march into Moscow, but what they found was a city abandoned and burned, denying France shelter or supplies.

In late October, after waiting too long, Napoléon decided to retreat to France. He was facing a winter that was especially harsh, even by Russian standards, at one point with a recorded temperature of 35 degrees below zero.[27] Most of his army had been injured, killed by war or weather, or had simply deserted. Nearly as important, almost 200,000 of his horses had died, severely limiting the speed Napoléon so often used to fight his battles.[28] There was little to counter the Russian Cossacks, expert horsemen who harassed the army throughout the retreat.

At one point, he left his army. He knew his only chance to hold off the advancing coalition would be to supplement his own troops, now down from the half million he took to Russia to some 10,000.[29] France, during the revolution and under Napoléon, was the first country to impose a large-scale conscription of its people.[30] He was able to bolster his army, but other countries joined Russia, forming a Sixth Coalition. Austria and Prussia, though they had signed treaties with France, switched allegiances and, of course, England, despite now being at war with the United States, supplied troops as well. There were some French victories, but France was fighting defensively now. There was a key loss at Leipzig, the bloodiest battle in history until World War I.[31] France was being invaded from the east by a combination of forces and from the south by the future Duke of Wellington. Talleyrand sought to end the fighting through negotiations, but also by passing military secrets on to the coalition.

France's defeat was imminent. With the call for more troops, higher taxes, war almost constant, and defeats now coming more often than victories, the French people increasingly desired an end to war. This Sixth Coalition was already making plans concerning what a postwar France would look like. Some wanted a return to a monarchy. (Louis XVI's brother was biding his time in England.) Some wanted Napoléon's current wife, Marie-Louise, to take over, acting as regent for her son. Some were even all right with a defanged Napoléon remaining in power.

Napoléon made an unsuccessful suicide attempt. He then tried to abdicate in favor of his son, but that was rejected. Finally, with his generals refusing to continue the fight, Napoléon unconditionally abdicated. The victors decided he would be sent away.

Napoléon was still allowed to rule, but on the island of Elba, not in France. In April 1814, Napoléon made the journey to Elba, brought there by a convoy of British ships. Neither his current nor former wife would be with him. Marie-Louise and their three-year-old son (who would die at age 21 from tuberculosis) returned to Vienna. Descended from Austrian royalty, Marie-Louise was given regions in Italy to rule over, which she would do for the rest of her life. (Napoléon's first wife, Josephine, passed away during Napoléon's exile.) His situation certainly could have been worse. Elba had a population of 12,000 and its residents welcomed their new emperor enthusiastically. Not exactly a prisoner, he was given an income, allowed a group of soldiers under his command, and was free to roam the island at will. He oversaw building projects, improved sanitation, revamped its legal system, and even designed a flag. It was not the empire he once led, but it was more than he might have had; it had an outcome, however, the victorious coalition would come to regret.

Ten months after his arrival, with only superficial oversight by the British stationed there, he simply sailed back to Toulon, the site of his first victory. He marched north, gaining popularity and troops as he went. Understandably alarmed, a Seventh Coalition was formed. Taking the offensive, Napoléon led troops north to meet them. They clashed just south of Waterloo, in what is now Belgium, where he faced British troops under Wellington and Prussian troops under General Gebhard von Blücher. Napoléon was able to defeat Blücher at Ligny and force his retreat, giving him the opportunity to concentrate on Wellington. June 18, 1815, would be Napoléon's final battle. Recent heavy rains limited his ability to move as quickly as he wanted. It was not until later in the day that he could finally get his artillery into position. At a crucial moment in the battle, Napoléon employed his Imperial Guard he had held in reserve, and they might have broken through, but Blücher's army dramatically reappeared at that same time and turned the tide. Defeated, the French troops fell back to France. Waterloo ended Napoléon's reign, but likely wasn't decisive. Other coalition troops were driving toward France at the time and even a French victory at Waterloo would likely have only forestalled the inevitable.

Again, Napoléon abdicated. St. Helena, his new home, would be far more bleak, remote, and heavily guarded than was the case on Elba. Elba had been just off the coast of France. St. Helena was in the middle of the Atlantic Ocean, a thousand miles off the coast of Africa. He was no longer an emperor, but a prisoner. He spent the rest of his life there until his death, likely from stomach cancer, on May 3, 1821.

VI

The Discovery of Egypt

Few regions of the world can rival the culture and history of Egypt. The Nile River is the world's longest river. Egypt held two of the Seven Wonders of the Ancient World. Even young children are aware of mummies, pyramids, the Sphinx, hieroglyphs, and names such as Nefertiti, Cleopatra, and King Tut.

Most scholars believe hieroglyphics to be one of the world's oldest written languages, second only to Sumerian. Some believe it may, in fact, be the oldest.[1] Those sacred writings of the Egyptians are more than 5,000 years old. Papyrus, the precursor to paper and nearly as old as hieroglyphs, was first made from the reeds along the banks of the Nile River. Egyptians developed a 12 month, 365 day per year calendar. The water clock and the sundial were Egyptian inventions used to record shorter periods of time. Egyptians were master builders, having constructed the pyramids which, until recent times, were the world's largest man-made objects. The building of the pyramids was based on their knowledge of astronomy. The base of the Great Pyramid at Giza points due north, south, east, and west to within a half degree of accuracy. The Egyptians also made advances in medicine. Though of little use today, the act of mummification demonstrates their knowledge of the human body.

The Egyptian civilization is one of the world's oldest. Greece is considered an ancient empire with its beginning in approximately 500 BCE. Yet, from the establishment of Greece to the present day is nearly the same amount of time as the years from the first Egyptian pharaoh to Greece's beginnings.

Alexandria was founded, given its name, and would be the burial place of Alexander the Great. It hosted the largest library of the ancient world and became a lure to the leading scholars of the day. A center of learning, it was home to scholars such as Euclid, Eratosthenes, and Archimedes. Euclid wrote *The Elements*, the world's most important text on early mathematics; Claudius Ptolemy (no relation to Egypt's pharaohs) wrote the *Almagest*, the most important text on astronomy for over a

115

thousand years; and Galen's writings were the most important in the field of medicine for, again, over a thousand years. Jewish scholars gathered in Alexandria to write the *Septuagint*, the Greek translation of the Hebrew Bible.

And yet, little was known of the land of Egypt before 1800 CE. Most travelers that did go there only went as far as its coastal cities. The language of ancient Egypt was dead and had not been understood by anyone for 1,500 years. The modern world became aware of Egypt when Napoléon Bonaparte landed there with an army of soldiers and scholars to conquer and explore. This not only began a process of releasing its secrets, but also created a whole area of study called Egyptology—the study of Egyptian history, language, and culture. It also launched a world-wide phenomenon that could be called Egyptomania—a fascination with all things Egyptian. Advertising, fashion, architecture, literature, and art all reflected this interest. Ancient Egyptian artifacts became prized possessions. There are now more Egyptian obelisks in Rome than there are in Egypt. When designing a monument to honor its first president, the United States chose to fashion it in the shape of an Egyptian obelisk.

Launching an Armada

General Napoléon Bonaparte's victorious Italian campaign made him a national hero. Following that success, the Directory called on him to lead an invasion of England. Bonaparte saw the difficulty of simply getting across the channel due to England's superior naval strength. He explained, "We shall not for some years gain naval supremacy. To invade England without that supremacy is the most daring and difficult task ever undertaken.... We must give up the expedition against England."[2] The newly named foreign minister, Charles-Maurice de Talleyrand, proposed an expedition to Egypt. If England were not invaded directly, control of Egypt would be a possible indirect attack. India, containing valuable resources, was now strictly a British colony after France lost its colonial presence there following the Seven Years' War. From England, there were two routes to India. There was the sea route around the southern tip of Africa, and there was the shorter route through the Mediterranean Sea, crossing the isthmus in Egypt, then sailing from there to India.

French possession of Egypt would disrupt that route and maybe even reestablish France's claim. Talleyrand, Bonaparte, and the Directory all came to see the advantage of this plan. It might also have been in Bonaparte's mind that two conquerors he greatly admired, Alexander the Great and Julius Caesar, had also made excursions to Egypt. From the

Directory's point of view, naming him the leader of this expedition held the added advantage that it moved the popular, maybe too popular, Napoléon Bonaparte away from France. If the mission were a success, all the better for the nation, and if it were a failure, that would take Bonaparte down a few notches. So, the Directory agreed.

An expedition would sail from southern France in May of 1798. Secrecy was the key, however. Bonaparte knew that he must get his army to Egypt by crossing the Mediterranean Sea while avoiding the British, the possessor of the world's greatest navy. To continue the ruse, the expedition kept its original name, The Army of England,[3] though it would soon be known as the Army of the Orient. Except for a select few, those going would not be told where they were headed until nearly at their destination.

When Bonaparte and Talleyrand first discussed the plan, they decided that if their Egyptian proposal were accepted by the Directory, Talleyrand would travel to Constantinople. There he would assure the Ottomans, the possessors of Egypt at the time, that they had no need for concern. He would let them know that the French expedition was not one of hostile intent against the Ottoman Empire and would not be a permanent occupation of the land. They only sought to dispatch the Mamelukes, who were ostensibly ruling on behalf of the Ottomans. In France's view, the Mamelukes were making a mess of things, including disrupting French trade. Throughout his diplomatic career, Talleyrand often proved to be disingenuous to better his own position. He and Bonaparte would spend their lives being dependent on, and suspicious of, one another. Such was the case here, as Talleyrand's promise to go to Constantinople proved hollow, and whatever his reasons were, he never made that contact with the Ottomans.[4] It is unclear why he didn't go, or if he ever even intended to go. Their surprise at the French invasion in what certainly seemed like an act of aggression, would eventually lead to their retaliation, and ultimately doom Bonaparte's expedition.

Roughly 50,000 soldiers, sailors, and scholars on 400 ships of various size would make the trip.[5] Accounts of the size of the expedition vary because Bonaparte would often fudge numbers if it was to his advantage to do so. In this case, he did not want the Directory to know he was taking more troops than were allotted. The armada embarked from several locations, including Toulon, the site of Bonaparte's first great military accomplishment, and Ajaccio, the city of his birth. Bonaparte, who was a part of the convoy of ships from Toulon, embarked on May 19, 1798, on board *L'Orient*, one of the largest ships in the world. They met up with the other ships along the way.

The scholars, known as savants, were brought to satisfy Bonaparte's desire to make this a scientific expedition as well as a military one. There were 167 savants possessing a variety of expertise.[6] Few of them knew what they were getting into when the excursion began. Violence or disease would ultimately cause 34 of them to die in Egypt.[7] There were cartographers, chemists, painters, musicians, astronomers, architects, and engineers. All the equipment they thought might be needed was loaded on board the ships. Scores of books covering the various disciplines were brought. There were telescopes, chemical supplies, survey equipment, transits, and ballooning equipment (along with thousands of bottles of wine). In a few weeks most of it would be at the bottom of the Mediterranean Sea.

Bonaparte spent a good deal of the voyage dealing with seasickness, but when not feeling poorly, he led discussions among the savants. These were nightly affairs meant to explore various scientific and philosophical topics. What is the nature of electricity? What is beyond the heavens? How old is the world and what is its ultimate end? Is there life elsewhere in the universe? Is there life after death? Bonaparte's intelligence and educational background allowed him to be on nearly equal footing with these learned men. Bonaparte's endeavor to make this expedition both a military and a scientific one is rare in the history of the world.

The fleet came to the island of Malta on June 10, approximately three weeks after leaving Toulon. Malta was occupied by the Order of the Knights Hospitaller of St. John. They were established as a hospital in Jerusalem during the First Crusade. Due to conflicts, including subsequent crusades, the Knights Hospitaller relocated several times, ultimately ending up on the island of Malta. By 1798, the Knights had inhabited the island for over 250 years. Bonaparte requested they be allowed to land to obtain fresh water. This was just a ploy, because his plan all along was to conquer the island. When not receiving the kind of welcome he was hoping for, Bonaparte proceeded with an invasion.

The Knights of Malta were fortified behind high walls, but the troops defending those walls were inexperienced and there simply were not enough of them to hold back the French troops. Seeing the inevitability of their defeat, the Knights surrendered within a day of the French landing. A treaty containing promises that Bonaparte had no intention of keeping was signed on board *L'Orient*. He went to work devising a new set of laws and having his team of savants make improvements to the island. He put an end to the Knights' monasteries, slavery, and feudal system. He freed its political prisoners. He established a postal system and a university. Bonaparte also helped himself to much of the wealth the Knights possessed. Gold and other treasures were loaded onto their ships, though most of this treasure would one day join the savants' equipment beneath the sea. Leaving a presence of 3,000 soldiers on the island, the armada and their newly won cargo were on their way to Egypt.[8]

It was only on this final leg of the trip that the crews were made aware of their ultimate destination. The armada arrived off the coast of Egypt, July 1, 1798. Because they were being hunted by the British navy, Bonaparte wanted to make landfall as soon as possible. The shallowness of the bay did not permit the larger boats to enter close to shore, so, even though the sea at the time was extremely rough, troops and supplies were shuttled there on smaller boats. With some loss of life and materiel, they landed, then marched through the night to reach the city of Alexandria, eight miles away.

Alexandria had lost its former glory and had a population approximately one-tenth what it once was.[9] The city was taken by the French troops with not much more difficulty than it took to capture Malta. There, the French saw their first ancient monuments—Pompey's Pillar and Cleopatra's Needles (neither of which had anything to do with their namesakes). Pompey's Pillar was standing and still stands there today. Cleopatra's Needles, dating from the fifteenth century BCE, were two obelisks, one standing and one fallen. Both are now in different locations. The fallen one was moved in 1871 and now is upright and sits on the bank of the Thames River in London. The standing one is in Central Park in New York City, having been moved there in 1880. Fair or not, the awakening of interest in Egypt following France's exploits led to many of its artifacts being moved to museums and various other locations throughout the world. Another obelisk, this one from the Temple of Luxor is now in the Place de La Concorde in Paris, on the site that previously held Paris' guillotine.

The Battle of the Pyramids

A week after taking Alexandria, Bonaparte organized his troops to advance to Cairo, Egypt's capital and largest city. General Jean-Baptiste Kléber, who had suffered a serious head injury in the attack on Alexandria, was left with a contingent to hold the city. The majority of the troops would head south, though without adequate maps. The maps they had gave barely more than an idea of where the major cities were. The cartographers attached to the French expedition would eventually develop maps that were much more helpful.

Cairo was located almost 150 miles south of Alexandria, and the path taken by the French troops was almost entirely through desert. The journey was extremely difficult. They marched in temperatures of nearly 120 degrees[10] while dressed in heavy woolen uniforms, soon leaving a trail of uniforms and equipment behind them. In the rush to organize the expedition, the men were not issued canteens,[11] and not nearly enough horses were brought. If they did come across a well or other source of water, it was likely brackish and nearly undrinkable. In previous campaigns, Bonaparte's armies fed themselves by living off the land as they traveled, but the desert region did not lend itself to this. There were other concerns that come along with desert life, such as snakes and scorpions. The sand made walking a chore, and it was extremely difficult to haul artillery. The blowing sand, along with the scorching sun, caused ophthalmia, a condition which led to severe eye problems, even to the point of blindness. On the way, soldiers were dying from thirst, exposure, exhaustion, even suicide, before reaching their goal. A Lieutenant Thurman later stated, "In the midst of a fine morning the atmosphere became darkened by a reddish haze, consisting of infinitely many tiny particles of burning dust. Soon we could barely see the disc of the sun. The unbearable wind dried our tongues, burned our eyelids, and induced an insatiable thirst. All sweating ceased, breathing became difficult, arms and legs became leaden with fatigue, and it was all but impossible even to speak."[12] At one point, the army came to the Nile River. Soldiers, ignoring orders from officers, ran and dove in. They found nearby melons, which they gorged on, making many sick, only adding to their woes.

As they neared Cairo, they came to their first real military test. The Mamelukes were the current power in Egypt. Centuries before, these Mamelukes (from the Arabic word for "slave") had been taken from the Caucasus region of Europe and trained for war. They were placed into the Muslim armies who were conquering much of the Middle East at the time. The Mamelukes had been in charge in Egypt until the Ottoman Empire invaded in 1517. Egypt became an Ottoman province with the Mamelukes allowed semi-autonomous rule of that area, paying an annual royalty to the Ottomans for that privilege. This money came out of the high taxes collected from the Egyptian people.

The two beys (warlords) in charge at this time were Murad Bey and Ibrahim Bey. They were prepared to face General Bonaparte's troops as they neared Cairo. Random Bedouins had been a nuisance along the way, picking off stragglers, and there was a skirmish with Mamelukes on July 13, but the real battle took place a week later. These two beys were in competition for control of the land but had an uneasy truce as they faced a common enemy in the French. Murad Bey faced them directly on the west side of the Nile River. Ibrahim Bey's people were on the far side of the Nile, but for the most part, were unarmed and did not take part in the battle.

Before the battle, Bonaparte told his troops, "Soldiers, 40 centuries of history are looking down on you."[13] He was referring to pyramids that were about ten miles away and not at all prominent on the horizon. In spite of the distance, he referred to this as "The Battle of the Pyramids" because he liked the sound of it. His estimate of 40 centuries was surprisingly close to what is now thought to be a span of 44 centuries.

The French were formed into battle squares. A battle square was a military formation that had been in use for many years but had recently seen a reemergence. Often more rectangle than square, sides were composed of two to six rows of soldiers. Bonaparte's squares contained roughly 4,000 men, each with artillery placed at the corners, with assorted baggage, generals, and cavalry inside. Battle squares were especially effective if there was enemy cavalry swirling around, and that is what

Murad Bey was a leader of the Mamelukes, a group serving the Ottoman Empire, overseeing their possession of Egypt. Though defeated by Napoléon's armies, his forces remained after France's had left.

they faced now. Some of the Mamelukes were riding, though many were on foot. The French had guns, while some Mamelukes had guns, but most were armed with scimitars, spears, and various other weapons, useful only for close-in combat.

The battle lasted two hours and consisted of several Mameluke charges. They were no match for French firepower and ultimately had to retreat with heavy losses. Many tried to escape by diving into the Nile, resulting in many of them drowning or, if not, being shot by French soldiers standing on the banks. Murad Bey headed south and would later be pursued by General Louis Desaix. Ibrahim Bey headed north toward Syria, where he and Napoléon Bonaparte would meet again.

Life in Cairo

Bonaparte's troops crossed the Nile and entered Cairo, a city containing approximately a quarter of a million people. Bonaparte thought that his troops would be seen as benevolent liberators, releasing the Egyptian people from the tyranny of the Mamelukes. Contrary to what Bonaparte had hoped, they were seen simply as invaders. As the French entered the city and settled in, the residents were in fear. To counter this, Bonaparte set about to extend the ideals of the Enlightenment to the land of Egypt. He established hospitals. Two newspapers were begun—one primarily for the savants and one for a more general reader. A postal system was established. He began the minting of money, provided street lighting, and cleaned up piles of garbage that had collected on the streets. A pontoon bridge was built which crossed the Nile River, the first bridge to ever do so. Despite these improvements, for the next three years, the Egyptian people would see the French as no less foreign intruders than they had the Mamelukes.

Egypt had its first ever printed books. Bonaparte had previously sent Gaspard Monge and Claude Berthollet to Italy to pick through the spoils of war after his victorious campaign there. Most of what they collected were works of art, ending up in the Louvre, but from the Vatican they also took printing presses. So, Egypt gained its first printing press—one that could print in Latin, Arabic, and Syriac script.[14] The newspapers and handbills posted throughout Cairo spread the word of the rules and regulations that were coming. Bonaparte established a rule of law for his own men and for the inhabitants of the city. He let merchants know the price of food (which he set), the new tax system, and various other changes.

A part of Bonaparte's plan of reconciliation was to convince the people of his support for their Muslim religion. He claimed to be a believer himself. He told them, "I respect God, his prophet Muhammad, and the Koran far more than the Mamelukes do."[15] However, he was probably being more honest when he said privately, "I find the idea of God is very useful to maintain good order, to keep men in the path of virtue and to keep them from crime."[16] He promoted the idea that he was the coming Messiah that was spoken of in the Koran. He even considered perhaps one day authoring a new set of scriptures. Bonaparte stated that he and his army wanted to convert to Islam, though this fell apart when it was learned that circumcision and abstinence from alcohol would be requirements.[17]

The people of Egypt typically celebrated Muhammad's birthday. Bonaparte was disappointed when he found out they were not planning a celebration this year due to the current occupation. Bonaparte insisted. The festival then took place with the French footing the bill and planning three days of festivities. There was music, dancing, fireworks, and parades. Later, Bonaparte would also make sure that the annual celebration of the Nile flooding its banks continued, and he also added a festival marking the anniversary of the French Revolution.

It is difficult to know what Bonaparte's religious beliefs truly were.

His declarations on the subject varied based on their usefulness to him at a particular time. For Bonaparte, the ends always justified the means. He professed a belief in Christ at certain times in his life, though in Egypt he stated, "The power of God passes through me so that I defeat the enemies of Islam and crush the Christian cross."[18] As he once said, "I always adopt the religion of the country I am in."[19]

He attempted to use the power of the church to strengthen himself politically—not always successfully. Pope Pius VI was opposed to the French Revolution in general and against the Civil Constitution of the Clergy in particular. In 1798, General Bonaparte had the pope arrested and taken to France, where he would die in captivity. In 1804, now head of the French government, Napoléon used Pope Pius VI's successor, Pius VII, to legitimize his standing with the French people. He invited, or otherwise convinced, the pope to travel from Rome to officiate at his coronation as emperor. It was held at the once again named Notre Dame Cathedral, with Napoléon having restored its original name after its revolutionary era name, The Temple of Reason. The pope's role, however, was that of a figurehead because Napoléon simply crowned himself. Pius VII later excommunicated the emperor for France's invasion of Rome, which he countered by arresting the pope, who remained imprisoned until the end of Napoléon's reign. While Napoléon's stated belief or disbelief in Christ varied, during the final weeks of his life on St. Helena, he attended weekly mass, and wrote in his will that he died in the Roman Catholic faith.[20]

Bonaparte walked a fine line between conciliation and repression. He instituted policies and made changes that improved the quality of life for the Egyptians, yet it was clear that the French meant to be in charge. The artillery stationed in the cities made that evident. The Egyptians were required to give up their firearms and were whipped or worse if they refused. Some of the policies or actions, which might have seemed to be in the people's best interest, were insulting to them and to their religious beliefs. Taxation, though less severe than under the Mamelukes, taxed religious sites. One of their mosques was converted to a tavern. Strategies to suppress the plague, such as quarantines, rules for times between a death and burial, or where bodies could be buried, were often an affront to the Egyptians.

The actions of French soldiers, and simply the fact that they were there, led to the Egyptians' animosity. Though Bonaparte set rules for his troops, along with severe punishments for violators, they often turned to thievery. The color green was to be worn only by the descendants of Muhammad, but, unknowingly, the soldiers often wore it.[21] Even the fact that the soldiers were clean-shaven was an insult to the Muslim men.

Egyptian women began emulating the French women they observed, and the Egyptian men were not happy about it. Some of the women even married French soldiers. Egyptian women, who usually stayed indoors, now went outside with heads uncovered and wearing brightly colored clothing. They began socializing with French soldiers, laughing and carrying on with them. They rode donkeys. All of these actions infuriated the Egyptian men.

In October, just three months after their arrival, the French were taken completely by surprise when a major uprising took place in Cairo. Speaking in Arabic, leaders were able to openly inform the people of the plans for the revolt. Weapons were surreptitiously distributed to the Egyptian men. From the French perspective, they were attacked in spite of all the improvements they had brought to Egypt. From the Egyptian perspective, the French who said they were there to set them free, were enslavers. During the revolt, Muslims who would not wear shoes in their mosques, now saw French soldiers riding their horses into those same mosques. In the mosques, soldiers trashed objects, including Korans. They drank the sacramental wine, then smashed the bottles.[22]

It took 12 hours, but the French regained control of the city. Three hundred French soldiers, including General Dominique Dupuy, Bonaparte's appointed governor of the city, were killed in the fighting. Five of the savants, scientists who had no idea what they were getting themselves into when they left France, were among the dead. On the Egyptian side, thousands were killed in the fighting. In the immediate aftermath, Bonaparte dealt severely with any insurrectionists that were found still in possession of firearms. They were rounded up, beheaded, and their bodies thrown into the Nile.[23] In the future, there would continue to be punishments, including beheadings, only increasing the animosity between the Egyptians and the French. While modern sensibilities might recoil at such actions, what the French did in the aftermath of battles was not necessarily worse than what the Mamelukes or other conquerors would have done.

A fragile peace was now in place. The uprising showed Bonaparte that his hope to be seen as a benevolent ruler of the Egyptians was not possible. Winning the people over was now transformed to more of a military occupation. A leading voice of the French Enlightenment, Jean-Jacques Rousseau, had felt primitive man was unspoiled. Man, in his primitive state was basically good. It was civilization that ruined him. Rousseau wrote of "the noble savage" as being pure and uncorrupted. Generally a believer in enlightenment principles, Bonaparte increasingly saw the natives as savages and anything but noble.

The French troops were growing disturbed at their situation. On edge after the riot, they also had to deal with the heat, sickness, and an epidemic of eye disease. Flies were abundant. While composing a letter, one of the savants wrote, "There are 12 flies on my hand as I write this."[24] Many of the officers requested permission to go back to France, sometimes asking their doctors to come up with medical reasons why they needed to return. Thomas-Alexandre Dumas was an accomplished general who had played an important part in the Italian campaign. He commanded the French cavalry in Egypt. However, he became increasingly critical of what was taking place, openly confronting Bonaparte. He was allowed to leave and sail back to France. However, his ship was caught in a storm and had to land in Italy where he was captured and imprisoned for two years before getting home. Years later, his son, Alexandre, would share his own commentary on French times in novels such as *The Three Musketeers* and *The Count of Monte Cristo*.[25]

Déodat de Dolomieu's fate was similar to that of Dumas. He was a famed geologist for whom the mineral dolomite and Italy's Dolomite Mountain range are named. He was a noble who, during the revolution, saw many of his friends sent to the guillotine, and himself nearly so. Like many of the savants, he was happy to be one invited to go to Egypt. As in the case of Dumas, he was allowed to leave Egypt after writing a report Bonaparte interpreted as being critical of him. On the same ship as Dumas, he was captured and held for almost two years.[26] Having been one of the Knights of Malta, he had assisted in the negotiations that led to Malta's surrender. It did not help his situation when it turned out that Knights of Malta were among his captors. He made good use of his time in prison, though, using the margins of a Bible and pieces of coal to write his important text, *The Philosophy of Mineralogy*.[27] Two main characters in Alexandre Dumas' *Count of Monte Cristo* seem to be based on the imprisonments of his father and Dolomieu.[28]

Part of the French soldiers' discontent came from the fact that, without a change in circumstances, they were now seemingly trapped in Egypt. That was because, not long after their arrival, the British Navy appeared and destroyed the ships that had brought them.

The Battle of Aboukir Bay

Admiral Horatio Nelson had the task of finding the French fleet, but so far, had been unable to do so. Though the French attempted secrecy, it was hard to completely hide hundreds of ships and tens of thousands of troops. Both England and France had spies in each other's countries, and Nelson soon learned where troops were gathering and had his own guess as to where they might be headed.

Horatio Nelson had been in the British navy since he was 12 years old and, by this time, was already a British naval hero, having given up an eye and an arm in the service of his country. Even greater fame awaited him because his upcoming naval battle with the French would make him Lord Nelson, and in seven years, though it would cost him his life, his victory at Trafalgar would make England undisputed master of the seas.

There were occasions where Nelson almost found the French fleet in the Mediterranean. He just missed it after it had departed from Toulon, and later unknowingly passed relatively close to it while at sea. Due to the French excursion to Malta, Nelson got to Egypt first. Not finding the French fleet there, he departed. The French arrived the day after the British fleet had left. Three weeks later, Nelson received word that the French indeed were in Egypt with their fleet anchored in Aboukir Bay, near the city of Alexandria. The French had 13 battleships led by *L'Orient*, the largest in the world, containing three decks and 120 guns. They were called ships of the line, simply because of the linear formation they took during battles. That is how the French ships were arrayed as Nelson approached.

Nelson made two strategic, though unorthodox, moves. He was able to squeeze some of his ships between the French line and the shore. He did this in spite of the fact he had no way of knowing the depth of the water. Though arriving in late afternoon, he prepared for battle. Due to visibility issues, sea battles were rarely fought at night, but Nelson began firing on the French just after six o'clock in the evening with the fighting continuing through the night. The French line was pounded from both sides. The French and the British had approximately the same number of ships and the same amount of firepower, but the British navy was simply better. The French navy had been hurt during the years of the revolution with many desertions and resignations, a lack of adequate training, and a general lack of discipline. Many of the navy's sailors were under 18 years of age. In addition, the French navy was undermanned at the time of the attack, having much of the crew ashore with the army troops.

The French fleet was commanded by Admiral François-Paul Brueys. He was part of the French navy that fought off the coast of America during its revolution. An aristocrat, he was arrested and demoted during the French Terror, but later reinstated. Now he was facing the British fleet from on board *L'Orient*, which was in the middle of his line of ships. Early in the fighting, he sustained a head wound, and later a cannon shot took off both of his legs. He had himself strapped to a chair and remained on deck, directing the battle from there for the next hour until he died.[29]

At approximately ten o'clock that night, *L'Orient* exploded. Being the largest ship of the fleet, it contained much of the scientific equipment, acquired treasures, and gunpowder. Under British bombardment, it caught fire and that fire spread to its store of gunpowder. The resulting explosion was deafening. Witnesses said that it was so loud it was heard in Alexandria, fifteen miles away.[30] Perhaps in shock, the combatants simply stopped fighting for a period of time.

The Battle of Aboukir Bay (sometimes geographically incorrectly called The Battle of the Nile) was an enormous British victory. Only two French ships of the line were able to escape. The others were either destroyed or captured. There was now no chance the French troops could return, nor could they be reinforced or resupplied. The fleet's commander, thousands of French sailors, supplies, and treasures were lost. The soldiers had lost their gunpowder; the savants, their scientific equipment. There were still many of the smaller ships, but they would have trouble accomplishing anything with the British in control of the Mediterranean Sea.

In the aftermath of the battle, Bonaparte unfairly laid the blame upon Admiral Brueys. His report to Paris detailed Brueys' supposed errors. According to Louis de Bourrienne, Bonaparte's private secretary and a personal acquaintance since their days together at the military academy at Brienne, unfairly ascribing blame to others was a common tactic of Bonaparte's. "The full truth was never to be found in Bonaparte's dispatches when that truth was even slightly unfavorable and when he was in a position to dissimulate. He was adept at disguising, altering, or suppressing it whenever possible. Frequently, he even changed the dispatches of others and then had them printed, whenever their view differed from his own or might cast some aspersion on his reputation and actions."[31]

Compounding his military losses were Bonaparte's personal issues. While his wife, Josephine, regularly found it difficult to be true to him, this was especially so while he was hundreds of miles away in Egypt. He had gotten reports of Josephine's dallying and wrote letters to her expressing his heartache. Communications could occasionally get through the British gauntlet, but often they were intercepted by enemy ships patrolling the Mediterranean. Such was the fate of several of Bonaparte's letters to his wife. In England, they were published in the local papers, to the great enjoyment of their readers.[32]

The Savants

Despite his troubles, Bonaparte saw to it that the work proceeded. Almost immediately upon their arrival in Cairo, Bonaparte set up what he called the Institut d'Egypt. A year earlier, he himself had been elected to the Institut de France and would fashion his Egyptian version after the French one. Several adjoining Egyptian palaces were used to create a science center and meeting area. The institute was organized into four areas:

physics, mathematics, political economy, and the arts. Bonaparte placed himself into the area of mathematics. He named Gaspard Monge president of the institute and himself as vice president. That Bonaparte held these countrymen in high esteem is clear, if overblown, when he said, "All men of genius, all those who have achieved distinction in the Republic of Science, are French, no matter what their native land."[33]

Gaspard Monge was an excellent teacher and administrator. He was instrumental in the creation of the École Polytechnique, still an elite institute of higher learning. He developed the area of mathematics known as descriptive geometry. This important field allows a three-dimensional object to be displayed two dimensionally—a valuable tool for anyone in a field of design. Descriptive geometry was recognized by the government as being so important that, for a time, it was treated as a military secret. He was a supporter of the revolution and then of Napoléon, but when Napoléon's time of importance came to an end, so did Monge's. When the monarchy was restored under Louis XVIII, he lost his teaching position, his pension, and other honors.[34] Another important French mathematician, working in the same field of geometry and one that studied under Monge, was Jean-Victor Poncelet. He founded the field known as projective geometry. Much of Poncelet's mathematical work was done in a Russian prison, having been captured while he was a part of Napoléon's invasion in 1812.

Bonaparte envisioned the institute as being a collection of intellectuals who would advance science and better the Egyptian society. Some tasks were assigned, yet the members were also free to choose their own areas of exploration. The savants tackled such issues as the advantages of using windmills versus watermills, making the water of the Nile drinkable, exploring how sand dunes were formed, and how to go about replacing what was lost in the Aboukir Bay naval disaster. How did Egyptian mummification work? What exactly was happening with the mirages that were seen by the troops on the journey between Alexandria and Cairo? The institute's advice was also sought as new judicial and educational systems were planned.

The task of making up for the items that sank to the bottom of the sea fell primarily to Nicolas Conté. Bonaparte put Conté in charge of the workshops that constructed these items. He was a genius when it came to manufacturing various needed parts and gadgets and would prove vital to the continued work of the army and the savants. There would be a need for replacement parts—for firearms, telescopes, windmills, printing presses, and more. Objects such as medical instruments, various tools, and cooking ovens would have to be made from scratch.

Nicolas-Jacques Conté was born into a farming family in Normandy,

France, in 1755. Early on, he was a well-respected painter, but also showed an ability to create objects out of the barest of raw materials. An example of his skill was his development of the pencil. Pencils, at the time of the revolution, contained pure graphite. It was in short supply, however, due to France's need to import the needed graphite and England's blockade of their ports. Conté developed a blend of clay and graphite and then heated it. By his process he was able to create pencils of varying hardness. He has been called the inventor of the modern pencil.[35]

During the revolution, the Committee of Public Safety endorsed the use of observation balloons to aid in the war effort. Conté and Jean-Marie Coutelle jointly led in their development. At that time, a common way of generating hydrogen gas was to use sulfuric acid. However, the committee did not allow its use for ballooning because it was reserved for the production of gunpowder, which was in short supply. Conté and Coutelle, in consultation with chemist Antoine Lavoisier, experimented to find other possibilities. As part of this experimentation, Conté felt he had discovered a varnish that could coat balloons, allowing virtually no air to escape. In the course of his experiments, a glass beaker shattered, causing Conté to lose the sight in his left eye.[36] He would wear a patch over that eye for the rest of his life.

Both men were part of the Egyptian team, and Bonaparte asked them to launch a balloon as part of a celebration. Most of their ballooning equipment was lost in Aboukir Bay, so they had to basically start from scratch. There were two flights, but they were more to impress the natives than as a part of scientific research.

In addition to improving conditions in Egypt, the savants' other task was to learn about the land they had taken over. This French expedition into Egypt has been called the beginning of modern archaeology.[37] There had been explorations into past civilizations, including in Egypt, but the efforts of these savants were much larger in scope and depth than anything that had been attempted before.

After the Egyptian campaign was over, an enormous set of books, called *Description de l'Égypte,* was published. These books described the land of Egypt and discoveries made by the savants. In 1802, Napoléon, by that time firmly in control of the French government, called for the publication of these books, and named Nicolas Conté to oversee the project. It would tell the world what Egypt was really like and be not only a guide for lay readers throughout the world but also for scientists' future studies. The first edition came out in 1809, but the entire set was not completed until 1829. Various editions ran from 20 to 30 volumes. The individual books were gigantic, roughly two and a half by three feet. Later, smaller versions, which were more affordable and more accessible to the general public, would do much to spark the world-wide interest in this mysterious land. Many members of the Institut d'Egypt contributed articles or sketches. In 1803, the ever-resourceful Conté created an engraving machine that was able to copy the many large drawings that were contained in the book. Without his machine, those sketches would have taken a great deal of time and effort if done by hand, assuming it could have been done at all.

The flora and fauna of Egypt were studied, drawn, and, when possible, taken back to the institute. Étienne Geoffroy Saint-Hilaire, noted naturalist, collected many animal specimens. At the institute, these specimens were then dissected, stuffed, or kept alive in a growing menagerie. Animals such as crocodiles, ibis, ostriches, flamingos, and camels were new, exotic creatures to these French adventurers.

Extensive drawings were made and measurements taken of the pyramids and temples. Many were the first sketches done by any European. An extensive topographic map was made. The map was felt to be so much better than any other in existence that it was kept as a military secret.[38] Astronomer Nicolas-August Nouet used his knowledge of the nighttime skies to set up a grid of latitude and longitude to aid in mapping.

As part of their collections, the savants housed a large block of granite that contained three different scripts. It was found in the ruins of the city of Rosetta. Immediately sensing its importance, those in charge had it sent directly to the institute. The savants could read the Greek portion, but the other two were indecipherable to them. There was a sense, though, that it could be the key to opening the mysterious Egyptian hieroglyphs. They were correct, although that would be years away.

Bonaparte joined his savants, taking part in the explorations. He visited the nearby pyramids, of which his "Battle of the Pyramids" was named. Among them was the Great Pyramid of Giza, the only surviving wonder of the Seven Wonders of the World. For thousands of years, it had been the tallest man-made structure in the world. Bonaparte, with his background in mathematics, calculated its bricks could be remade into a wall surrounding France that was nine feet high and three feet thick. Savant Edme-François Jomard made extensive measurements and calculations, using trigonometry to find additional heights and angles. Bonaparte and others watched as Gaspard Monge climbed to the top of the Great Pyramid. This was especially impressive because while in France, Monge had tried to beg off going on the expedition at all since, at 53 years of age, he felt he might be too old for such things. He only went after Bonaparte went to his wife to get her to join him in convincing Gaspard to go.

They visited the Sphinx, approximately a quarter mile from the Great Pyramid. The Sphinx was buried nearly up to its neck due to centuries of blowing sand. Years later, it would take the efforts of two French men to uncover it. The Egyptian Council of Antiquities formed to stop the illegal trade in Egyptian artifacts. French scholars took a leading role in this organization its first hundred years. (Ironically, it was France's expedition to Egypt that created the demand for Egyptian artifacts, and thus the illegal trade in them.) In 1858, Auguste Mariette, the council's first director, had some of the sand removed that was surrounding the Sphinx. In the early 1900s, the council had engineer Emile Baraize finish the job of excavation. Though the sand removal was an immense project, the sand likely protected the Sphinx through the centuries, preserving its appearance. Napoléon Bonaparte would not have seen much more than the Sphinx' head sticking out of the ground. (The story that it was missing its nose because his French soldiers used it for target practice is incorrect. The nose likely had already been missing for centuries. It was a common practice for enemies to deliberately remove the nose of a statue as a sign of disrespect.)

In December of 1798, accompanied by soldiers and several savants, Bonaparte journeyed south. In part, he wanted to check on French forts that had been established, but he also sought to find the route of a canal that had been built by Ramses II. The current Suez Canal connects the Mediterranean and the Red Sea. Ramses' canal was in a different location, stretching from the Nile River to the Red Sea. They found the remnants of it and followed its path many miles until it was simply swallowed up into the desert. Bonaparte considered building a canal that would form the link between the Red and Mediterranean seas. One of the surveyors found that there was a 33-foot difference in height of the two seas.[39] Breaching that difference would call for something like the system of locks currently used in the Panama Canal. It turned out that the surveyor's measurements were incorrect. The heights were the same, and a canal could have been constructed without the use of locks. Regardless, upcoming events would have kept the canal from being built by Bonaparte's men. It would be another 70 years before the French, under the guidance of Ferdinand de Lesseps, built the Suez Canal.

Bonaparte sought to see biblical sights on his journey. Though he did not achieve his goal of seeing Mt. Sinai, he did see, and even crossed, the Red Sea. Following were events reminiscent of the book of Exodus. He and other savants crossed when the tide was low. They later became disoriented and did not make the return trip until night had fallen, by which time the tide had risen. Bonaparte and his party were nearly swept away and killed. The horses they were riding had to start swimming when the water had risen to their necks. "We were only too happy not to have shared the fate of Pharaoh's soldiers."[40]

Denon's Travels on the Nile

While there were many savants working in various areas, perhaps the most important contributions were made by Vivant Denon. He accompanied French troops as they traveled up and down much of the Nile River in search of Murad Bey.

Vivant Denon was born in 1747 in Charlon-sur-Saône, located in the middle of France. He was privately educated, then in his twenties moved to Paris where he studied law and began learning the art of engraving. A capable young man, he held several positions in the royal government. He was appointed by Louis XV as the keeper of gems that Louis had inherited from his mistress, Madame de Pompadour. Denon went on to hold a variety of diplomatic positions in the years prior to the Revolution until he was requested to travel to Egypt as one of the savants. At 51, which he referred to as his "advanced age," he decided to go.[41]

He was part of the expedition as it crossed the Mediterranean and then marched on Alexandria and Cairo. After the occupation of Cairo, General Jacques-François Menou was sent north and Denon went with him. Menou, who would soon convert to Islam, take an Egyptian wife, and change his name to Abdullah, was made the governor of Rosetta by Bonaparte. Able to explore the delta region, Denon described finding a room that contained arms and armor that apparently had been undisturbed since the time of the Crusades, 500 years before.[42]

Denon and other savants, mostly artists and archaeologists, accompanied General Desaix and over 3,000 soldiers on a nearly year-long expedition along the Nile, stretching to the southern edge of Egypt, covering over 600 miles. Louis Desaix was 30 years old at the time of the Egyptian expedition. Prior to this, he had led troops in the Austrian campaign, serving with Napoléon Bonaparte. After the return from Egypt, he would again fight in Austria against the Second Coalition assembled against France. There, he would be shot and killed during battle.

Desaix was sent up the Nile River in pursuit of the Mameluke army. Although Murad Bey was repulsed at the Battle of the Pyramids, he had fled south and was still in possession of a formidable fighting force. After their failed frontal assault at the Battle of the Pyramids, Murad was pursued and would be dealt another blow by General Desaix. From that point on, Murad shifted to a guerrilla campaign, enlisting the local Egyptians to rise up against the French. Additionally, the French troops were faced with random attacks from tribes of Bedouins, along with the ongoing annoyance of bandits taking any opportunity to make off with whatever they could steal.

While Denon and his fellow savants would later bring Egypt to life for a Europe which knew little of the land, there were many difficulties in their efforts. Though Desaix was generally supportive of the savants' work, his primary goal was to track down Murad Bey. That meant Denon was not able to spend nearly the amount of time he wanted at some of the locations. Sometimes it meant recording his thoughts or making sketches of passing ruins as he marched by or sat atop a camel. Denon often wrote of his explorations being cut short by rocks, spears, or gunfire aimed his way. In addition to the threat of violence, there were harsh physical conditions. Though the journey generally followed the Nile, Desaix sometimes ventured into the desert if he got word that Murad was in a nearby area. Denon spoke of winds that arose in the desert, leading to tremendous sandstorms that could turn day to night. They were surrounded by, as he described, all kinds of vermin. He and a companion once woke in the middle of the night with red, inflamed skin from receiving bites as they slept. In spite of traveling along the Nile for most of the journey, several died due to the heat.

And yet, while the hardships were many, Denon was thrilled when he came to a new find. "Pencil in hand, I passed from object to object, drawn from one by interest in the next, constantly enthralled, constantly distracted; my eyes, my hand, my mind were inadequate to the task of ordering and setting down all that overwhelmed me."[43] He made over 200 drawings during his travels south.

Denon was alternately greatly impressed or disappointed by the condition of the temples and artifacts he encountered. Many buildings, both their exteriors and interiors, were in excellent condition because the arid climate enabled a state of preserva-

Vivant Denon traveled with the French army as it traveled the Nile. Denon's future writings and drawings would do much to popularize Egypt with the outside world.

tion. And while thieves likely had already stolen anything of monetary value years before, for the most part, what was of primary value to the savants remained. Denon often commented that the paintings on the walls appeared as brilliant and colorful as if they were done just prior to his viewing them.

On one occasion he found a piece of papyrus in the hand of an Egyptian corpse. Even in a dry climate, items such as papyrus would decay over time; however, in cases like this, and like that of the Dead Sea Scrolls, papyrus can stay in good condition for thousands of years. Denon carefully took the papyrus and shared it with the institute when he arrived back in Cairo. Excitedly, the savants attempted to make out the writing, but with no success. That piece of papyrus would not be deciphered until Jean-François Champollion did so 20 years later. It turned out to be a portion of the Egyptian Book of the Dead; a collection of spells which aided an individual's journey through the afterlife.

Many buildings, paintings, and artifacts, however, did not age well. Though not a major area of seismic activity, over time, earthquakes had toppled some of the temples. With an increase in population and industry, pollution has faded the writing and the art, especially in the 200 years since the French expedition. The obelisks that have been transported to cities such as Paris and New York have remained standing, but their hieroglyphic writing has almost faded from view. One reason Denon's writing is so important is because he described places and items that no longer exist today.

Thieves had stolen much of worth. The rooms within a pyramid containing the mummy of a pharaoh also contained gold and jewels. Much that was of monetary value was already gone before the French arrived. Even the resin used in mummification had been stolen and resold in Cairo. Builders of the pyramids sought to put barriers inside the pathways leading to those rooms. Thieves often simply tunneled through the softer surrounding limestone to get to the treasures. Though most of the precious metals and jewels were gone, because of the French expedition and the resulting interest in Egypt, thieves soon found an increased market for stolen antiquities.

Denon and the other savants got to visit many sites on their journey, though depending on the army's itinerary, some sites were quick visits or simply bypassed altogether. Typically, they were able to spend at least some time at a valued location going either up or down the river. They saw sites rich in history, including The Valley of the Kings, Karnak, Luxor, Aswan, Philae, and Thebes. There were rich discoveries in every city visited. Abydos was one of the oldest sites. Many of the earliest pharaohs were entombed there, and it became a center of worship of the gods Isis and

Osiris. Much of Abydos now lies under modern buildings. Faiyen was a city that worshipped a crocodile named Sobek. When the Greeks came, they gave the city the descriptive name, Crocodilopolis. Aswan showed examples of Egypt's varied history all in one location. There he saw Greek columns in the Doric style, Roman Emperor Trajan's Kiosk, and the foundations of a Catholic Church.

Other savants visited Antioöpolis, a city that was established by Emperor Hadrian in 130 CE. Most Roman emperors spent their reigns within the borders of Italy. Hadrian, though, ventured forth, exploring much of the world. Little of the city remains today. Over the last two centuries, much of it has been razed and used as material for new building projects. When Denon visited it, though, he could write of seeing its streets that were lined with colonnades on each side, along with its theater, hippodrome, and temples.

At Tentyris, the temple of Hathor contained the Dendera Zodiac. Now a treasured artifact, the world was not aware of its existence until Denon's visit. Not just showing signs of the Zodiac, it displayed a map of the sky. It was theorized (wrongly) that it showed a map of the heavens, including both a solar and lunar eclipse, in what was calculated to have been the year 50 BCE. Denon made a sketch as best he could. That sketch would later assist François Champollion with his translation work. The roof containing the Zodiac was later removed by antiquities dealers with a variety of saws and gunpowder. It is now displayed in the Louvre Museum.

Denon was only able to spend three hours in the Valley of the Kings but said that he could have spent three days. The Valley of the Kings contains many tombs of pharaohs. Now one of the world's most important archaeological sites, until this journey it was unknown to the world. Of the eight tombs he visited, he described six as being in a state of perfection, with the other two having only some water damage. He said the walls looked as if they had been freshly painted.

Exploring a chamber was often a struggle. With no source of light, Denon had to do his explorations by torchlight. He had to take breaks to go outside and get fresh air, and he sometimes competed with scores of bats living inside. Despite these sometimes difficult circumstances, he spent hours copying the images he found inside. The mummies he found were not limited to those of humans. The ibis was a bird sacred to the Egyptians, representing Thoth, the god of wisdom. More plentiful in ancient Egypt than now, its image was often pictured on temple walls and inside pyramids. Denon once found the mummified remains of over 500 ibis in one location. Millions more of their mummies have been found since. There are only a few living ibis left in the world today, and none in Egypt.

Denon was able to observe Egypt's irrigation technology. The Nile flooded annually, depositing water and nutrient-rich silt to the land on its borders. Ditches and canals extended its reach even farther. As they traveled, the savants took measurements of the river's height. At one point, the Nile was rising at two inches per day, but days later had increased to a rate of 12 inches per day.[44] At Qena, they found the Nile was six feet higher on the return trip than when they passed it originally.[45] They examined what were called nilometers. There were various styles of nilometers, but a typical one was a large well made of bricks and containing a staircase spiraling downward. One could walk down the steps to read the Nile's height on an attached ruler. The ruler could be attached to the wall or might be a vertical pole running down the center of the well. The scale would have been marked off in cubits, a cubit being the distance from elbow to the tips of fingers. This information was used by the Egyptians to know when earthworks should be broken, releasing the Nile's water onto fields. Information on how much water would be available in a flood season was also used to determine tax rates.

Ancient Egyptian astronomers also tracked flood cycles by watching the Dog Star. Part of the constellation Canis Major (*Canis*, Latin for "dog"), the Dog Star, was invisible for a part of the year due to its nearness to the sun. It became visible again in late summer. This coincided with the flooding of the Nile. The late summer months when the flooding took place are still referred to as "the dog days." These early Egyptian astronomers also used the heavens to forecast various other seasonal events. Flooding of the Nile River, however, along with the use of nilometers, is now a thing of the past since the building of the Aswan Dam in the 1960s.

After a journey of well over a thousand miles, the expedition arrived back in Cairo. Denon made a presentation to the institute and one personally to Napoléon Bonaparte. Savants tried to translate his copies of hieroglyphic symbols but were unable to do so.

Denon was soon to leave Egypt. He would be among a select handful chosen to secretly accompany Napoléon Bonaparte to France. In France, Denon would continue to play an important role. Bonaparte appointed him to direct the Louvre Museum, which he led for ten years. However, Bonaparte's fall would lead to Denon's fall as well, losing his position at the museum in 1815. He collected his writings and sketches into a book entitled, *Voyage dans la Basse et la Haute Egypte* (Travels in Upper and Lower Egypt). It was published in Paris in 1802, London in 1803, and in many other editions and languages. Its popularity and influence were enormous and did much to spark a world-wide enchantment with Egypt. The original size was massive, with some of his illustrations stretched over two pages. Rather than making such sketches by hand, he was aided by his own invention of a machine that could be used for engraving. Originally, the books could only be afforded by the rich, but later editions were smaller and more affordable. His book was a major reason for the world's coming fascination with Egypt.

Syrian Debacle

Egypt had become somewhat pacified. The Cairo rebellion had been put down, Ibrahim Bey had retreated to the north and Murad Bey, while not disposed of, was greatly diminished in strength. However, Bonaparte had gotten word that trouble was brewing in Syria. The Ottoman Empire, England, and Russia had formed an unlikely alliance and were gathering in Syria, an element of a second coalition that was preparing to take on the French. Rather than wait, Bonaparte decided to go on the offensive. He led 13,000 troops north in February of 1799.[46] After the army's experience in the desert, lighter uniforms were made. However, being February on the Mediterranean coast, there would be the opposite problem of the troops suffering from the cold.

Aside from meeting the immediate threat, it is not clear exactly what Bonaparte's goals were. At various times, he spoke with confidence of defeating those troops gathered in Syria, then perhaps pressing on to take Constantinople, the center of the Ottoman empire. After being victorious there, he might return to Egypt or perhaps simply continue on through Europe, attacking the formed Second Coalition, and finally arriving triumphantly in Paris. Or perhaps, after victory in Syria, he might turn east and head toward India as Alexander had done two millennia before. Perhaps his ultimate goal varied by the day.

The first encounter of real consequence was at El Arish, which surrendered after a short siege. Bonaparte had no use for taking prisoners who would need to be guarded and fed. He allowed those captured to leave with their promise they would not take up arms against the French for a full year.

Proceeding up the coast, the army came to Jaffa. A peace emissary was sent with the hope that the city would realize its fate and surrender rather than troubling each side with a siege. In what turned out to be a bad idea, the leader of Jaffa did not simply deny the request but beheaded the emissary and placed his head on a pike, displaying it for the French to view.

After a three-day siege, the French army charged through a breach in the city walls. The French victory was complete, with several thousand killed or taken prisoner. Bonaparte was unhappy with the great number that were allowed to surrender. He was again faced with the issue of what to do with them. Was he to somehow transport them back to Egypt, drag them along on his way north, or release them as he had done at El Arish? Their fate was sealed when he found that maybe half of the prisoners were those prisoners he had just magnanimously released at the previous siege. He had them put to death.[47] The prisoners were taken to the shore of the Mediterranean, where French soldiers were commanded to shoot them, though they were later told to use bayonets to save ammunition. Their bodies were thrown into the sea. When General Berthier protested the decision to kill them, Bonaparte suggested he join a monastery.

The citizens of Jaffa were undergoing an outbreak of the plague. In what might be seen as divine justice, the French soldiers then suffered their own outbreak after taking the city. Though the French had been dealing with the plague since first arriving in Egypt, it was to become an epidemic among the troops during this expedition north.

Bonaparte had brought a cadre of doctors and nurses on the Egyptian campaign. René-Nicolas Desgenettes was the chief physician, and Dominique-Jean Larrey was the chief surgeon. Both were accomplished and experienced physicians. Politically, Desgenettes was a Girondin during the revolution, and thus, at odds with the more powerful Montagnards. He left Paris after the fall of the Girondins and accompanied Napoléon to Egypt to serve as a surgeon. In Cairo, after the Syrian campaign, Desgenettes dramatically confronted Bonaparte at a meeting of the institute. Bonaparte had laid blame on the medical corps for not containing the plague. Desgenettes not only confronted him about his accusations against the medical team, but also his leadership. He also accused Bonaparte of ordering him to poison those with the plague so they would not have to be transported. (Testimony varies as to whether this actually happened, and if it did, whether this was a heinous act or one of compassion.) In the future, he would serve the French troops in European battles in the ill-fated Russian campaign and at Waterloo.

Doctor Larrey, as well, served in European battles, including Waterloo. He has been called the first modern military surgeon, setting up hospital units near the front lines of battles. These were forerunners of later wars' MASH units. Different types of ambulances had been in existence but were used to collect the wounded after a battle was over. Larrey created what were known as flying ambulances.[48] These were horse-drawn vehicles that were used to triage, treat, and transport the wounded as the battle transpired. The injured were then delivered to hospitals behind the front lines in stabilized condition to receive further care.

The physician René Laënnec was one of those that was serving in hospitals behind the front lines. In 1812, he was in charge of the wards in the Salprêtière Hospital, a facility reserved for wounded soldiers returning from the fighting. Laënnec did important research, including forever changing medical care with his invention of the stethoscope.

The medical obstacles faced in Egypt were many—the plague, dysentery, and ophthalmia were prevalent. The world at that time did not realize the plague was transmitted by fleas carried on rats or on a person's clothing. Without that knowledge, they did the best they knew, stressing isolation, cleanliness, and a variety of medicines. For anyone contacting the plague, it was likely a death sentence. Fear spread among the troops. That fear led one of the doctors to refuse to treat those with the plague. Bonaparte placed him in jail, but not before forcing him to ride a donkey through their encampment while wearing a dress.[49]

The French troops continued north, arriving at Acre on March 18. Ahmed Pasha el-Djezzar was the governor of Acre and the surrounding area and had been so for the past 25 years. He had the disturbing, but well-deserved, nickname "The Butcher." Infractions committed by his own people could be punished with the loss of an ear, nose, or hand. During the siege of Acre, he made a habit of cutting off French heads and displaying them on the city walls.

Aside from the naval battle at Aboukir Bay, in which he did not have a direct role, Acre was Bonaparte's first military defeat. The French surrounded the city by land and assumed the sea would complete the encirclement. However, Commodore Sidney Smith of the British navy was offshore, having arrived just a couple of days before the French. Smith supplied the city with food and ammunition. French ships that were bringing siege guns from Egypt were captured by Smith; not only were the French deprived of their firepower, but Smith then used those weapons against them.

A member of Smith's crew was Louis-Edmond Phélippeaux. He and Bonaparte were classmates at military school. They had an adversarial relationship at that time, and it was now about to get much worse. After his schooling, Phélippeaux commanded a French regiment, but his royalist leanings were at odds with the politics of the day, and he emigrated to England. He later returned to France to be part of a royalist insurrection. The insurrection failed and he was imprisoned. He narrowly avoided the guillotine, managing an escape the day before his scheduled execution. He met up with Commodore Smith and joined his fleet. At Acre, he used his military school training to improve the city's defenses by building a moat and trenches, strengthening city walls, and then mounting guns on them.

Bonaparte had brought some siege guns overland with him and a few also made it past the British sea blockade, but there were not enough to have nearly the firepower that was needed. The typical siege strategies did not work. With the lack of firepower, bombarding the walls was ineffective, and the city could not be starved because they were being resupplied by the British fleet offshore. Several attempts to storm the defenses failed, being repulsed each time.

General Louis-Marie-Joseph Cafferelli, a favorite of Bonaparte, was among those in the fight. He already had a wooden leg, a result of fighting the First Coalition in Europe in 1795. On one occasion, while Cafferelli was trying to inspire his troops, one of his soldiers supposedly quipped, "Easy for you, you still have one foot in France."[50] In Acre, he was hit by cannon fire, shattering his right arm. Doctor Larrey amputated what was left of his arm, but gangrene set in, and he died days later.

Bonaparte had enough. Using a strategy others had employed before and since, he simply declared victory and left. He issued a proclamation congratulating his troops, of which his secretary Bourrienne stated, "This proclamation, beginning to end, mutilated the truth."[51] The French spiked their siege guns and left in the middle of the night. By the time they got back to Cairo, they were approximately two-thirds their original number.[52] In a foreshadowing of his Russian campaign, the army made a difficult, inglorious retreat. The winter cold they felt as they came north was now replaced by the intense dessert heat of June. Dysentery, eye disease, the plague, and caring for the wounded made the month-long trip back an ordeal.

When they did reach Cairo, Bonaparte made sure they had a hero's welcome. Accompanied by stirring songs of a band, they marched through a triumphal arch. He sent a message to France stating, "I razed Djezzar's Palace to the ground, along with the ramparts of Acre. There is not a stone left standing, and all the inhabitants have left by sea. Djezzar is seriously wounded."[53] All lies.

There was a period of calm in Egypt, but it was interrupted a month later when British and Russian ships transporting Ottoman troops landed on the shores of Aboukir Bay. They set up fortifications on a peninsula jutting into the Mediterranean. Bonaparte mobilized his army and achieved a great victory, literally pushing the opposing forces into the sea. Bonaparte himself led the charge. While there were many reasons why his troops might hate him for what he put them through in Egypt, it was this bravery and willingness to share in his men's burdens that endeared him to them.

Things were not going well in Egypt, but they were not going well back home either. A second coalition had formed against France. Many of the previous military gains had been wiped out and France's borders were now being threatened. The Directory wanted Bonaparte to return, and seeing no clear path to victory in Egypt, he must have been only too happy to comply. As a scientific pursuit, the expedition to Egypt was a great success, but militarily, for the most part, it had been an utter failure.

Bonaparte told virtually no one of his plans. He left with the savants Monge, Denon, Berthollet, and enough military men to safely make the trip. That was all. They sailed away in secret, hugging the coastline to avoid detection by the British. Luckily, Smith had sailed to Cyprus to replenish supplies, allowing Napoléon's crew a safe passage home. Forty-seven days later, he landed in the south of France. His journey north to Paris was one of universal acclaim. And why wouldn't it be? France was in danger of being overrun by England, Russia, and Austria. And now here was the second coming of the one who saved France from the First Coalition. Additionally, he was, according to the reports he sent from Egypt, coming home from a victorious Egyptian campaign.

Meanwhile, though the fighting in Egypt continued, the end was in sight. More enemy troops had landed and more had marched overland from Syria. General Kléber was now in charge, having found this out in a letter that Bonaparte had left behind for him. A few months later, while walking in a garden in Cairo, he was stabbed to death by a 24-year-old Arab fanatic. (Another of Napoléon's generals, General Desaix, who had pursued Murad Bey up the Nile, would die that same day, fighting alongside Napoléon in Italy at the battle of Marengo.) General Menou took over and soon negotiations for a French surrender began, though the French had very little bargaining power. The savants wanted desperately to keep all their work and threatened to destroy it all rather than surrender it to the British. Ultimately, the French were allowed to keep the papers and specimens they had collected. However, this agreement did not include any items determined by the British to be antiquities. This included the Rosetta Stone, which, even at that time, was considered a prized possession. It was shipped to, and remains in, London's British Museum. Over the years, a person's perspective as to the fairness of the negotiated treaty has generally been aligned with that person's nationality, the British feeling it more than fair, and the French, not.

Napoléon's secretiveness with actual numbers makes it difficult to determine the exact count, but it appears that of those that went to Egypt, over half died there, with many others coming back with serious injuries.[54] The French troops had managed to hang on a couple of years after Bonaparte's departure, returning home in the fall of 1801. Thus ended the three-year Egyptian campaign.

In France, Emmanuel Sieyès, a former priest, was making plans to topple the unpopular and ineffective Directory. For a successful coup, he believed he needed to be associated with some type of figurehead—someone popular that the people would rally behind. He would also need that person to be someone that could be moved aside after a successful coup, allowing Sieyès to be fully in control.

This coup, known as the Coup of 18 Brumaire, was very messy, but was ultimately successful. A new government was headed by three consuls consisting of Sieyès, Bonaparte, and Roger Ducos. However, Sieyès' choice of someone that would passively relinquish power was quite wrong. Bonaparte was not that type. He soon managed to have himself named first consul. Bonaparte was becoming Napoléon. The Revolutionary Era was over. The Napoléonic Era had begun.

VII

Unlocking the Hieroglyphs

It is difficult to underestimate the influence of Egypt on the world. Egypt was an important land in history and had been on the world stage for millennia. Sitting on the corners of Asia, Europe, and Africa, it was in a vital location. It invaded or was invaded by the Assyrians, Babylonians, Israelites, Persians, Greeks, and Romans. It was one of the earliest, if not the very earliest, in the development of a written language, paper, and a calendar, as well as establishing work in the fields of astronomy, writing, geometry, surveying, and medicine.

However, for all the work done by the savants, there was still so much to learn about Egypt. Little was known of ancient Egypt, and a good deal of what was supposedly known was wrong. The history of Greece and Rome was well documented because the languages of Greek and Latin had been read and studied by scholars. Much of what could be learned of Egypt was contained in its ancient writings—writings which the world could not understand or even pronounce. This left an enormous gap in the knowledge of world civilization.

Egypt had been visited by Julius Caesar, Marc Antony, and other Roman leaders, and became part of the Roman Empire under Caesar Augustus in 30 BCE. It would remain so for another 600 years. When Emperor Constantine declared Christianity to be the official religion of the Roman Empire in the fourth century CE, there was no longer room in the empire for what they considered to be pagan religions. Egypt's religion and its pagan rites, such as mummification, would have to go. Its temples were turned into churches. Hieroglyphs (literally, sacred carvings) soon became a dead language.

The world in 1800, including among the Egyptians themselves, did not know the history of this important land. Egyptian history was lost, despite the fact that its writings were displayed in front of the savants on the walls of temples and pyramids. While the Egyptians believed that a pyramid would send a king or queen on a journey to the afterlife, they also believed that could only happen if his or her name and the dates of his or her reign were written on its walls.[1] A person's name was considered a living thing, and a mummy could not be reinhabited without it. The history of many of these leaders of ancient Egypt, even their very names, would be unknown today without the ability to read the hieroglyphic writings.

The last known hieroglyphic writing was on a temple on the island of Philae in southern Egypt. Once hieroglyphs were able to be translated, in the nineteenth century, this temple was found to have a date written on its walls corresponding to August 24, 394 CE.[2] (The island of Philae is now underwater, due to the building of the Aswan Dam. In the late 1970s, the temple was taken apart and reassembled on higher ground.)

The hieroglyphic Egyptian language, however, would prove especially difficult to translate. It was so difficult that there is still translation work going on today. There are many reasons for this. Languages undergo changes over time. When seen by Napoléon Bonaparte's savants, hieroglyphs had not been in use for over a thousand years. Not only that, but they had been in use for the previous 47 centuries. Considerable changes have taken place in the English language since the days of Chaucer or Shakespeare who wrote a mere handful of centuries ago. Similarly, translators were trying to conquer a moving target with the Egyptian language that undoubtably had undergone changes over the course of that much time.

There were different languages and scripts in use throughout Egyptian history. Hieroglyphs were the domain of the priests, and they were among the few that could read them. A simplified form of hieroglyphic writing was known as hieratic. On the Rosetta Stone, along with hieroglyphs, there were the two other writings—demotic and Greek. There was also the Coptic language, still in use in the Coptic Church today, and a key to hieroglyphs' ultimate decipherment. All of these played a role in ancient Egypt until giving way to Arabic, the language of modern Egypt.

The Egyptian language, like all languages, had certain quirks, though Egyptian seemed to have more than most. Verbs came at the start of sentences. There were some semi-vowels, but for the most part, vowels were not used. There were no gaps between words. The writing could be displayed horizontally or vertically. If horizontal, sometimes it would be read left to right and sometimes right to left. The key would eventually be found to be which way any hieroglyphic animals were facing. If a sentence contained a hieroglyph displaying animals' profiles that faced left, the reading of its sentence would begin on the left and be read left to right. If the animals faced right, sentences would begin on the right and flow right to left. In the interest of symmetry, lines might alternate directions. Typically, the writing was horizontal, but depending on the available space, it might be most convenient to write vertically. If so, the writing would be top to bottom, but if there were several columns, the same ordering principle was used. If animals faced left, reading would start with the leftmost column. If facing right, the rightmost column.

The English alphabet has 26 letters. Russian has 32. There are hundreds of hieroglyphs. For those that sought to translate the language, the greatest stumbling block was determining what categories of language these symbols even represented—were they words, concepts, letters of an alphabet? English is alphabetic. Aside from the occasional numeral, emoji, or some other symbol, the English language is made up of 26 letters. Letters, grouped together, form words, and words strung together form sentences, the foundation of both written and spoken communication. English letters are not meant to look like anything in particular, but these hieroglyphs did. There were recognizable objects such as plants, owls, camels, and stars. It was believed that they could possibly stand for letters, but these objects did not look like letters.

For centuries, the common assumption was that hieroglyphs were ideograms—pictures of things or ideas. If so, a picture of a star could represent the word "star." A picture of legs could mean "legs" or "running" or "walking." All that was left was to determine the correct identification of the pictures. The problem was that the would-be translators were only partially correct. Hieroglyphs turned out to be a mixture with some symbols being alphabetic characters, some ideograms, and some could be either type, depending on the context. As Jean-François Champollion wrote after he successfully broke the code, "Hieroglyphic writing is a complex system, a script all at once figurative, symbolic, and phonetic, in one and the same text, in one and the same sentence, and I might even venture, in one and the same word."[3]

The Rosetta Stone

A key to eventual success would be found in a slab of stone from the Egyptian city of Rashid, or by its English name, Rosetta. It held three scripts, all of them containing basically the same message. The bottom portion was written in Greek. Even though it was ancient Greek, not identical to modern, it could be read by the French savants. An especially intriguing line was one that stated, "This decree shall be inscribed on a stela of granite, in the writing of the divine words, the writing of documents, and the writing of the Ionians." The "divine words" were the top section, hieroglyphs. This section was followed by "the writing of documents," which would be known to them as demotic, and "the writing of the Ionians," which was the Greek section.

It was quickly seen that the words in Greek at the bottom of the slab could possibly lead to translating the words in the messages above. One of the difficulties was that the three sections were not identical messages, but only close paraphrases of each other. The Rosetta Stone would be extremely important, but even with it, unlocking the secrets of hieroglyphs was many years away.

General Menou and French troops were stationed in Rosetta, which was one of the more beautiful areas of Egypt at that time. It was a coastal city 30 miles east of Alexandria in the fertile Nile River delta in an area filled with palm trees and other lush growth. Menou had his men working on building a fort. Because there was very little stone in the delta region, it was a common practice, even before the appearance of the French troops, to reuse stone from old buildings. The remains of a fifteenth-century Ottoman fort were being reused to build a new one for the French. It was in this reclamation process that Lieutenant Pierre-François Brouchard came across a large flat stone with writing on one side. Brouchard had trained to be a balloonist but had suffered an eye injury in an explosion. (It was the same accident that had caused Nicolas Conté to lose his eye.) In Rosetta, he was serving as an officer in the military, and, because of his engineering background, was also a member of the Institut d'Egypte. After the Egyptian campaign, he would continue to serve in the French army. He was taken prisoner by the British, but later was released. He would again side with Napoléon during his triumphant, but short-lived, Hundred Day campaign between Elba and Waterloo.

Upon finding the slab, Brouchard notified General Menou who, recognizing its significance, had it sent south to the savants at the institute in Cairo. In a couple of years, Menou would be the head of all the troops in Egypt, taking over for the assassinated Kléber. By 1801, the British had defeated the French, and Menou and the French savants were trying to salvage whatever they could of their scientific work. The savants obviously wanted, and felt entitled to, the discoveries they had made the past three years. The British did not see it that way, and the French were not in a position to change their view of things. In the end, the savants were allowed to keep their papers and items such as animal specimens, but anything the British considered to be an antiquity would stay in their possession. That included the Rosetta Stone. This was not entirely unreasonable considering the spoils the French had taken from other countries in the past.

Menou did all he could to keep the Rosetta Stone. He claimed it as his personal property; he hid it in his tent; he secretly moved it to Alexandria where it was hidden in a warehouse—all to no avail. The savants had copies made of its writings before giving it up. They used inks and wax casts to make usable, though imperfect, copies. This caused a blackening of the stone, which was only recently thoroughly cleaned and restored to its original color. The British loaded it onto a cart, drove it to the Mediterranean shore, and sailed to England, where it found its permanent home in the British Museum. On its side, still slightly visible today, was written, "Captured in Egypt by the British Army 1801. Presented by King George III." There it stayed, except when it was moved to a safer, underground location during the bombing raids of the world wars, and in 1972 when it was loaned to France to celebrate the 150th anniversary of Champollion's breakthrough discoveries.[4]

It was large, two and a half feet wide and four feet high.[5] The writings it contained would prove of great value to those trying to translate hieroglyphics, however, it would have been more helpful had the messages been complete. Its dimensions today are the same as they were when found 200 years ago, but portions had broken away in the years prior. The middle demotic has all of its lines, though is broken along the right side. The top section of hieroglyphics and the bottom section of Greek each had complete lines that were missing.

Upon the death of Alexander the Great, his conquests were fought over, with the land of Egypt going to Ptolemy, one of Alexander's generals. He was known as Ptolemy I, with many more Ptolemaic pharaohs to follow. They would rule Egypt for over 300 years, until the last Ptolemy, Cleopatra, committed suicide and Egypt came under the control of Rome. When he was five years old, Ptolemy V became pharaoh upon the death of his father, and at the age of 14, it was he that had the message carved onto the Rosetta Stone. The carving has a date on it which corresponds in the modern system to March 27, 196 BCE.

There was unrest during his reign. However, Ptolemy V had recently achieved a victory over a segment of rebelling Egyptians, and the carving on the Rosetta Stone was in response to that victory. There was lingering tension between Ptolemy V and the priesthood as to who was really in charge. The stone Ptolemy had made asserted that he was in charge, but made concessions to the priests to insure their agreement with that claim. Among other considerations, the priests received a reduction in their taxes, and the requirement that priests journey to Alexandria once a year was dropped.

A Confusion of Tongues

Part of what made the translation work so difficult was the variety of forms the Egyptian language took. There were two different Egyptian scripts on the Rosetta Stone, but there were others not included, which were ultimately important in its translation. The Greek portion of the Rosetta Stone was not simply a different script, but a completely different language from the others.

The savants recognized the hieroglyphic top section—it was all over the monuments of Egypt, though they did not understand any of it. The middle, called demotic, was a complete mystery to them. The bottom was Greek and was the part they could understand, although there was some difficulty due to their lack of familiarity with Egyptian culture and history.

While writing in Egypt used various media such as stone, papyrus, or pottery, hieroglyphs were almost always written onto stone. It was meant to last. The hieroglyphs might be painted, carved into the stone, or a bas-relief, in which the figures stood out from the stone. It was a language of the priesthood and had great religious significance, honoring the deities. The gods such as Ra, Isis, and Osiris were divine, as were the pharaohs, and even certain animals such as ibis, crocodiles, and cheetahs.

Hieratic was another Egyptian script, though not one written onto the Rosetta Stone. Both were considered divine writings (it and hieroglyphs shared the Greek root *hier*, meaning sacred) and only understood by the priests. Hieratic was, however, more easily written. Only the hieroglyphs would adorn temples, pyramids, or obelisks. Hieratic could be used by priests to write about divine matters on media such as papyrus. While not one of the Rosetta Stone scripts, it would prove valuable to those seeking to be translators.

Over time, hieroglyphs evolved into hieratic, which evolved into demotic, the middle section of the Rosetta Stone. Hieratic made its appearance in approximately 3000 BCE and demotic in approximately 600 BCE. Sacred writing was not appropriate for the day-to-day life of the people, which is what led to the development of demotic. Demotic was a script to be used by the common people, sharing its root with words such as "democracy." With no need for permanence, the demotic script was typically written on papyrus, and thus, had mostly disintegrated by the time the French came to Egypt. Like hieratic, it was a shorthand, cursive version of hieroglyphs, and like the others, it would fade from use and was a language long-dead by the time the Rosetta Stone was discovered. That there was similarity between these three scripts was not at all obvious, and it was this discovery by the Englishman Thomas Young that would be a major breakthrough. Champollion later said the same. This would become one of the points that would lead to Young's claim of plagiarism by Champollion.[6]

Coptic differs in a number of ways from the previous three scripts. Coptic is still in use today, though those who understand it number only in the hundreds worldwide. Today, it is primarily used in Coptic Church services in Egypt. It gained its popularity in the fourth century CE when Christianity became the primary religion throughout the Roman Empire. It remained in common use until approximately the seventh century, when it was replaced by Arabic, which remains the language of Egypt.[7] The other Egyptian scripts were partially alphabetic, but Coptic is totally so. It was made up of 24 letters from the Greek alphabet but included six additional letters added to account for sounds made in the Egyptian, but not the Greek, language.[8] Vowels, for the most part missing from hieroglyphics, were a part of Coptic and aided in pronunciation. While the written form of Coptic is quite different from the others in appearance, Champollion would find that pronunciation of words in modern Coptic was similar to that of ancient Egyptian. It would be a key to his future success. He could hear the ancient Egyptian language before he could read it.

Greek comprised the bottom section. The previous four were different scripts, but all the same language. Greek was a completely different language. It was the official language of the government at the time of its writing on the Rosetta Stone. Alexander the Great, the conqueror of Egypt, and those who were in charge after, the Ptolemys, were all Greek. There was hieroglyphic, hieratic, demotic, Coptic, and Greek. Three were on the Rosetta Stone, but all five were part of Egyptian history and all would play a role in deciphering Egypt's original language.

The Rosetta Stone was a valuable find and started many attempts to translate Egyptian. However, the Rosetta Stone had limitations. The top and bottom sections had at least some damage, and though a search was made of the area where the slab was first discovered, those missing pieces were never found. What was left was not pristine. That was not surprising since it had been sitting among rubble for centuries. Many hieroglyphs are intricate and having even slight damage can make them difficult to correctly identify. A dog, a standing wolf, and a wolf lying down could be quite similar in appearance yet have different meanings. Only the British, and who they chose to allow viewing privileges (with the French likely being at the bottom of the list), had access to the original. Others, at best, had imperfect copies.

There were, though, sources beside the Rosetta Stone, but they had their problems as well. Most of the monuments were still in existence, although as time went by, thievery and decay would continue to reduce their number. Champollion had helpful documents in Paris, brought back after the Napoléonic campaigns, although after Napoléon's fall, many were returned to their original locations. As if not confusing enough already, because of the phenomenon of Egyptomania and an appreciation of its artistic value, some started using attractive but meaningless hieroglyphic-like symbols as decoration. Much of the act of translating was not simply the translation work itself, but also the search for reliable source material that could be starting points for the work to follow.

Thomas Young

Thomas Young was born in the small town of Milverton, in southwest England. Seventeen years older than Champollion, he was born into what would be a family of ten children. Apparently, his gift for language came early. He read by the age of two, and at four years old he had read through the Bible twice.[9] He attended schools in Scotland, Germany, and finally Cambridge University in England. His training was in the field of medicine, but his studies would encompass so much more.

Science has become specialized. However, there was a time in which scientists could do important work in a great many areas. Young's discoveries were varied and profound. He coined phrases such as "Indo-European" and "kinetic energy," his experiments helped to verify the wave theory of light.[10] He did research into such disparate areas as life insurance, acoustics, bridge construction, elasticity, and the tuning of musical instruments.

Thomas Young was a genius. Though unsuccessful in solving the puzzle of the hieroglyphs, his efforts were key in Champollion's ultimate discovery.

Young made important discoveries into the mystery of sight. He found that the eye can focus on objects at different distances by a warping of its lens, and not changes within the eye itself, as previously thought. He found that individuals see different colors because of red, green, and blue receptor cones in the eye. He was the first to describe astigmatisms.[11]

Though with an intentional exaggeration, author Andrew Robinson, titled his biography of Thomas Young, "The Last Man Who Knew Everything." In 1931, Albert

Einstein, writing in the foreword to Isaac Newton's *Opticks*, said of Young, "Reflexion, refraction, the formation of images by lenses, the mode of operation of the eye, the spectral composition and recomposition of the different kinds of light, the invention of the reflecting telescope, the first foundations of colour theory, the elementary theory of the rainbow pass us by in procession, and finally come his of the colours of thin films as the origin of the next great theoretical advance, which had to await, over a hundred years, the coming of Thomas Young."

Young and Champollion were very different people. Thomas Young grew up with parents that were strong Quakers. His faith seemed to be a major influence throughout his life though as an adult he strayed from his parent's beliefs just a bit, playing cards and attending dances. He would not use sugar because of its association with the slave trade.[12] Champollion was anti-clerical as were many others coming out of the French Enlightenment. Young was even-keeled, not given to highs and lows. Champollion was not even-keeled. He took offense if he felt others, which would come to include Thomas Young, diminished his work in any way. Young was not political. Champollion, on the other hand, wanted to see the power of royalty diminished, if not eliminated. He would come to be a supporter of Napoléon Bonaparte and would face severe consequences for that support.

While they shared a quest to unlock the Rosetta Stone, their motivations differed. Young was a polymath studying a wide array of topics. His venturing into hieroglyphics was one of his many areas of research, perhaps examining it only because it was a puzzle to be solved. He saw Greek and Roman societies as being important in the history of the world, he but did not believe this to be true of Egypt. As he said, "The great mass of Egyptian monuments of all kinds relates exclusively to the religious and superstitious rites observed towards the ridiculous deities and the idolized heroes of the country."[13]

Champollion was focused on the single area of linguistic scholarship, and especially on the study of the ancient Egyptian language. Champollion was not interested simply in the language, but also every aspect of the Egyptian world. Egypt had been a source of fascination since childhood. Its history, culture, and language would be the central concern of his life, and his expertise in these areas would be helpful to him in his translation efforts.

Champollion

Jean-François Champollion was born December 23, 1790, in Figeac, France, at the time of the French Revolution. While Figeac was not the hotbed of activity that some other locations in France were, the city guillotine was located only yards from the family home. Thomas Young was financially comfortable, but Jean-François grew up poor and would remain so for much of his life.

His father was a bookseller. This access to books would be helpful to Jean-François' education. However, his father spent most of his life in debt and addicted to alcohol. On one occasion, the adult Jean-François was living at home, and a bailiff came to collect an overdue debt owed by the father. Champollion paid the debt out of his own money. Later, expressing his frustration to his older brother, he wrote of having to cover for "our big baby of a father."[14]

His mother was likely illiterate. In ill health for many years, she passed away when Jean-François was 16 years old. His father was often away on business and was dysfunctional when at home. Jacques-Joseph, an older brother by 12 years, effectively became Jean-François' guardian, becoming his younger brother's encourager, tutor, and source of financial support. Jacques-Joseph's own education was affected negatively by the revolution. The Catholic Church operated many of the French schools but were closed for years due to the revolution's distaste for religious organizations. Even

after religious orders were allowed to reopen schools, many had had their land confiscated and sold to finance the war effort. Jacques-Joseph would continue his education largely on his own. He would go on to have an impressive career in his own right, serving in several governmental and university positions. However, sharing his brother's interest in Egypt, he would have great regret that he was not chosen to accompany Napoléon on his Egyptian campaign.

Jacques-Joseph would play an important role in his younger brother's education and overall well-being. He tutored Jean-François, then set him up with other tutors

Jean-François Champollion unlocked the key to translating the hieroglyphs of Egypt, unlocking thousands of years of history.

and finally enrolled him in schools when those became available. Thanks to him, Jean-François' education in language began early in life. A private tutor began teaching him Greek and Latin. As his education progressed, he would learn Hebrew, Chaldean, Ethiopian, Coptic, Persian, and Italian, among others.[15]

At the age of 10, Jean-François joined his brother in Grenoble where he was working at the time and when he was 12, Jacques-Joseph financed his enrollment in a private school, and two years later, in one of France's public schools. The public secondary schools, *lycées*, were established by Napoléon Bonaparte and were part of his reform of France's educational system. The *lycées* often had strict discipline and a militaristic tone that did not appeal at all to Jean-François. He disliked the school's harsh discipline, uniforms, and marching drills. After much pleading, Jean-François was finally able to convince his brother to take him out of the school and allow him to continue his language studies in Paris.

Though his time at the *lycée* in Grenoble proved trying, Jean-François would continue to learn about languages and cultures, though much of the learning was self-taught because those subjects were not a substantial part of the school's curriculum. At the end of his schooling at the *lycée*, he made an impressive presentation to the local Society of Arts and Sciences, which he entitled *Essay on the Geographical Description of Egypt before the Conquest of Cambyses*. It was well received and demonstrated his work ethic and ability to do important research.[16]

Health issues began to plague Jean-François during his school years. He was in poor health, often severe, throughout his life, and would die at a young age. He began life with a difficult birth. He possessed a driven, type A personality which would aid his work, but perhaps caused some of his physical difficulties. As he grew older, he gained a large amount of weight. For whatever reason, or combination of reasons, he dealt with many ailments which included breathing difficulties, extreme headaches, gout, insomnia, fainting spells, exhaustion, fever, and stomach and chest pains.

Jean-François studied in Paris for two years when opportunity opened for him and his brother. A university was located in Grenoble. Jacques-Joseph was hired as a professor of Greek literature, and Jean-François as an assistant professor of ancient history—a remarkable honor for someone not quite 20 years old.

In the summer of 1815, while the brothers were living in Grenoble and teaching at the university, they would encounter Napoléon Bonaparte. Napoléon was in the process of a slow, devastating retreat from Russia as the Sixth Coalition chased him back toward France. The brothers were a part of the defense of Grenoble as Austrian troops approached the town. Fortunately for them and the town, Napoléon abdicated before direct combat became necessary. He was banished to the island of Elba, only 30 miles from Corsica, the island of his birth. He was not considered a prisoner there. There was some British military presence, but more to keep an eye on things than to keep Napoléon a captive. The allies allowed him to rule a small island, and he quickly began building and organizing his new domain, but it was nothing like the glories of his previous life. At some point, he began planning for a triumphal return to France.

Louis XVIII was in place as the king of France. He was in the line of Louis XV and younger brother of Louis XVI, thus next in line for the throne. (Though never king, the son of Louis XVI and Marie Antoinette died in prison, but was recognized as Louis XVII.) However, there were still many that were not in favor of a monarchy and still saw Napoléon Bonaparte as a hero and the legitimate ruler of France. He was able to make his escape, though not an escape so much as he simply left. Napoléon and a small number of troops boarded a ship and sailed from Elba, landing three days later in the south of France. This began what is commonly referred to as The Hundred Days in which he made his final stand on the world stage. He marched north to Paris, collecting troops and supporters along the way.

Louis left the city as Napoléon's army approached, effectively handing the country over to him. Napoléon would spend the next several weeks organizing his government. It is not clear, perhaps not even to Napoléon, if he would be content to only rule France or if he would seek to acquire other country's lands as had been his habit in the past. The other countries were not willing to wait to find out. Predictably, a coalition led by England formed to deal with the perceived threat. It would be a seventh and final coalition. Choosing to take the offensive, he mobilized his own troops and marched north toward Belgium and the small village of Waterloo.

The Champollion brothers' contact with Napoléon came early in The Hundred Days campaign. Napoléon had enjoyed a growing celebration of France's citizens as he marched from the southern coast to Paris. A regiment out of Grenoble confronted him as he approached the city. In a dramatic show, he stepped forward, unbuttoned his coat and proclaimed, "I am here. Kill your emperor if you wish."[17] Apparently, that was enough to sway them to his side. He entered the city to great acclaim, and the Champollion brothers joined in the celebration. Napoléon stayed in the city for a time, and on the recommendation of the mayor of Grenoble, met with Jacques-Joseph.[18] Napoléon was impressed by him, enough that he appointed him to be his secretary. Jacques-Joseph traveled north with the caravan, arriving with them in Paris, and working from there until the toppling of the short-lived regime.

His younger brother was also able to meet with Napoléon. Since they both possessed a fascination with Egypt, the two of them spoke of discoveries that had been made, as well as Jean-François' own translation work in hieroglyphics. Napoléon told him he would see to it that a Coptic dictionary Jean-François had been working on would be printed in Paris (though future events would disrupt those plans).[19] After Napoléon marched northward, Jean-François remained in Grenoble where he edited a newspaper which was basically a pro–Napoléon propaganda journal. In that newspaper, he had written, "The people alone award the crown; they gave it in the past to Hugo Capet [Louis XVI] and now they have taken it away from his descendants so as to entrust it to someone more worthy. Their choice confers sole legitimacy. Napoléon is therefore our legitimate prince."[20] Unfortunately, and unbeknownst to him, it was written the same day the battle of Waterloo was taking place.

This time, Napoléon would not abdicate soon enough for Grenoble to escape invasion. Jacques-Joseph was in Paris at the time, but his brother fought along with the rest of the city as they faced Austrian and Sardinian troops.[21] With heavy shelling killing or wounding hundreds of its citizens, and Napoléon's time clearly coming to a close, the city surrendered.

In the aftermath, the brothers would face the consequences of being on the losing side. They were brought before the authorities. A few years earlier, under the influence of Robespierre, they doubtless would have faced execution. This time, although they were not officially charged, they lost their jobs in Grenoble and were banished from the town. They returned to the town of their birth, Figeac, where they moved back into the house they grew up in. The result certainly could have been worse, though it did mean living in a town without nearly the opportunities or resources for scholarship that they were used to. After almost a year, the older brother, and then the younger, were permitted to leave. Lessons were not always easily learned, however, for headstrong Jean-François. He would continue to voice his anti-royalist feelings and would later be tried for treason, though he would be acquitted.[22]

Initial Attempts

Many scholars sought to learn the secrets contained in hieroglyphs. While Young and Champollion stand out, there were several others working with the hopes of achieving a breakthrough.

Silvestre de Sacy was one of the leading linguists of the day, specializing in the languages of the Near East. He taught many of those who would go on to make important advances in the study of the Rosetta Stone, including Étienne Marc Quatramère, Johan Åkerblad, and Jean-François

Champollion. De Sacy was an initial encourager of Champollion's research, but they had a troubled relationship through the years. A major source of this conflict was likely that de Sacy was deeply religious and a royalist, and Jean-François was neither. De Sacy did not support the usurper Napoléon, and he vacated Paris during The Hundred Days of his return. Though there were times that he wavered, Champollion typically counted himself a republican and a supporter of Napoléon Bonaparte. De Sacy was, at times, critical of Champollion's theories, and on several occasions accused him of plagiarism. In a letter written to Thomas Young, de Sacy stated, "I would suggest you do not communicate too many of your discoveries to m[onsieur] Champollion. It could happen that he might afterwards lay claim to the priority. He seeks, in many parts of his book, to make it believed that he has discovered many words of the Egyptian inscription from Rosetta. I am afraid this is mere charlatanism."[23] There may have been some professional jealousy as well, since de Sacy had tried, but failed, to make much progress interpreting the Rosetta Stone. Though contentious in life, they seemed to have made amends at the end, with de Sacy acknowledging Champollion's discoveries and delivering a kind eulogy when he passed away.

Champollion's relationship with Silvestre de Sacy was complicated, but he would have other complicated relationships as well. Edme-François Jomard was likely his most severe and enduring critic. Jomard was a geographer and one of the savants on Napoléon's expedition to Egypt. He would later be named chief editor of the *Description de l'Égypte*, the massive, multi-volume undertaking that detailed the savants' explorations and findings. Nicolas Conté was initially in charge of the project but passed away in 1805. Then his successor, Michel-Ange Lancret, passed away in 1807. Jomard spent the next decade on this task, organizing the contributions of dozens of scientists and artists. The worldwide fascination with Egypt would not have taken place without this work. However, some of the information in the text that dealt with hieroglyphs was privately criticized by Champollion. These comments, and others made by him regarding Jomard's own translation attempts, got back to him and sparked the animosity. Jomard, in turn, privately criticized Champollion's efforts and lobbied to keep him out of academic societies to which he had applied. Jomard would continue his campaign against him even after his theories came into general acceptance by others, and even after his death.

Champollion also had differences with Étienne Marc Quatremère. At the height of The Terror, ten-year-old Étienne and his mother went into hiding after his father's execution, supposedly for treason. He would gain a professorship at the university at Rouen, France, and later take de Sacy's university post after his death. He was an early proponent of the value of

Coptic in revealing hieroglyph's secrets. Champollion found out Quatramère was to publish a book that had elements similar to what he was working on. Champollion rushed to be first to publish, but came in second to Quatramère. When he ultimately did publish, some, including de Sacy, at least hinted at plagiarism on Champollion's part. There probably was nothing to the charge, but this added Quatremère to his growing list of those to whom he felt bitterness.

Joseph Fourier was one of the world's most important mathematicians, and vital to the future success of Champollion. He had trained to become a priest, though that ended with the new revolutionary government's suppression of religious schools. Fourier became a proponent of the ideals of the revolution yet was opposed to the excesses during the Reign of Terror. He was arrested, but avoided execution only after Robespierre's death, and the subsequent diminishing of those excesses. He was chosen to be a part of the Egyptian expedition and was named secretary of the Institut d'Egypte. There, he was placed in charge of administering programs among the local people. Traveling in Egypt 30 years later, Champollion would be pleased to find Egyptians that remembered Fourier and what they felt was his fairness in his dealing with them. After their return to France, Napoléon appointed him to be prefect, or overseer, of the Department of Isère. He would stay in that position for 13 years until Napoléon's fall. He was based in Grenoble, which, at the time, happened to be the residence of the Champollion brothers. One of his duties was to inspect Napoléon's newly established *lycées* of the area. On one occasion, Fourier visited the school at Grenoble and heard presentations given by the school's top students. Jean-François was one of those top students, presenting on a portion of Hebrew text from the Bible. Fourier was impressed.

Through the years, he did much to further Jean-François' career. He was unhappy at the school and was given permission by Fourier to leave his dormitory and study from home. Desperately needing to bolster his armies, Napoléon instituted a draft, and on two separate occasions, Fourier intervened directly to Napoléon to keep Jean-François from being conscripted. Fourier was able to get him access to source material, including an improved copy of the Rosetta Stone. He gave Jean-François connections that enabled him to learn Coptic. Among those connections was that of a Coptic priest, Yuhanna Chiftichi. Jean-François studied with him and attended his Coptic Church services in Paris. Chiftichi was able to gain access to Coptic documents that had been taken from the Vatican and were now located in Paris.

Champollion and Fourier were part of an ongoing debate over the Dendera Zodiac. It was found and written about by Napoléon's savants during their time in Egypt. Years later it was taken from its temple home by adventurers and was now resting in Paris. Fourier felt it might be dated as far back as 15,000 BCE.[24] This was based on his opinion that the Dendera Zodiac displayed a map of the heavens that placed the time of its creation. This troubled church leaders since it seemed to contradict biblical accounts. Jean-Baptiste Biot, of science and ballooning fame, argued for a much later date. This controversy would finally be solved by Champollion. The source of most of the confusion came from the fact that the piece seemed to have both astronomical and astrological elements. Champollion felt it had more to do with astrology than being something that could accurately be dated by looking at its celestial arrangements. After unlocking its hieroglyphic secrets, he was able to date it to a time that fit much more closely to the church's position. The pope was so pleased that he offered to make him a cardinal.

There was another small glitch in their relationship. Fourier wrote an extensive preface to the *Description de l'Égypte* detailing the history of the land. Both of the Champollion brothers did research in preparation for the writing. They, however, were chagrined when it came out in print, and they saw that they received no recognition by Fourier for their efforts. In spite of these difficulties, theirs was a supportive relationship and furthered Champollion's research.

It was during his time in Grenoble that Fourier began extensive research into the nature of heat. He measured heat and its movement within a substance or its transfer between substances. His text, *The Analytical Theory of Heat*, is one of the important texts in the history of science. One consequence of his work was that he determined the earth's temperature was warmer than would be expected. He stated that the atmosphere acted as a blanket, keeping in the heat from the sun, a phenomenon that is now referred to as the greenhouse effect.

Many of his discoveries came about because of his development of what came to be known as the Fourier series. He found that any mathematical function could be rewritten as a sum of trigonometric functions. The advantages of doing so are not obvious, but profound. It is difficult to overstate the importance of his work. The Fourier series finds applications today in areas such as image processing, cell phone technology, quantum mechanics, and signal processing.

Breakthroughs

Silvestre de Sacy was one of the first to attempt to translate the language of the Rosetta Stone. He had some success in the demotic section, finding the names Alexander and Ptolemy.[25] He worked alongside his former student Johan Åkerblad, a Swedish diplomat whose work would go on to surpass de Sacy's, finding names such as Cleopatra, Berenice, Arsinoe, and Alexandria along with several words that were not proper names.

More than simply finding the translation of individual words, Åkerblad began to make discoveries regarding the nature of demotic and Coptic writings. His finds, though sometimes flawed, would set a basis for future work. He noticed a similarity between those scripts; a similarity Thomas Young would later expand on. He found proper names contained in the demotic portion and developed a demotic alphabet of 29 letters, although roughly half of those letters would prove to be incorrect.[26] Both de Sacy and Åkerblad felt that demotic was made up entirely of alphabetic letters rather than symbols standing for complete words, as was conjectured at the time. Thomas Young would later make the discovery that demotic was partially alphabetic, but also contained symbols that stood for words and phrases.

At least partially due to the Napoléonic Wars, progress stalled for several years until the work of Thomas Young and Jean-François Champollion, and they are the ones who would have most of the later success. Theirs was a mixed relationship, at times bitter and at times cordial. It began coincidentally with Champollion seeking a better copy of the Rosetta Stone. At a time when their two nations were not at war, he wrote to England's Royal Society. His request happened to come to, and was answered by, the Royal Society's secretary at the time, Thomas Young.[27]

Debate still exists as to how much credit is due each man, with alliances being drawn primarily along nationalistic lines. As with Leibniz and Newton, Edison and Tesla, and Lavoisier and Priestley, there probably will never be a resolution. The issue is confused because Young often published his work anonymously, and both men were guilty of leaving less than thorough notes on their work. Credit is also difficult to assign because each had major mistakes mixed in with major discoveries. They exchanged pleasant, helpful letters through the years and even met and worked together at times, but there were times of bitterness as well. Champollion once wrote, "The discoveries of Dr. Young, announced with so much splendor, are only a ridiculous bragging. The much praised discovery of the claimed key fills me with pity."[28] Direct confrontation would have been considered in bad taste, but there was plenty of opportunity for complaints to be shared privately to others.

Thomas Young was able to complete translation of the demotic portion of the Rosetta Stone. Focus now turned to the hieroglyphic section. Young found that the hieroglyphic and the demotic were not different languages, but different scripts of the same language.[29] He found that, historically, hieroglyphic gave way to hieratic, which then led to demotic.[30]

The differences were much more pronounced but could be considered similar to writing by printing verses in cursive. Young initially agreed with Åkerblad that demotic was based on an alphabet, but he later stated that demotic was made up of a mixture—partly letters of an alphabet and partly symbols for words.[31] A modern-day comparison might be the English language being made up of a combination of emojis standing in for words and phrases, along with a 26-letter alphabet.

While he believed hieroglyphs were symbols, he thought he saw an exception in what the French called cartouches. Unique to hieroglyphs; cartouches were oval shapes that surrounded groups of hieroglyphic symbols. Cartouches, so-called by the French soldiers in Egypt because they resembled the bullets (called *cartouches* in French) used in their guns, were theorized to contain foreign proper names. They were thought to show proper names in the same way that other languages mark them by capitalizing the first letter. How could there be a hieroglyphic symbol for words that were foreign to them? Without a symbol to use for the name, it would have to be spelled out in letters. A proper name such as "Roosevelt" is similar in every language. If a speaker of a foreign language were making a presentation containing the word "Roosevelt," that would likely be the only word the audience would understand. Modern-day sign language employs hand signs for concepts, but also possesses an alphabet for spelling items such as words for which there was no symbol. Such must be the case with foreign proper names.

In 1818, Young found "Ptolemy" in a cartouche on the Rosetta Stone. Though Ptolemy was an Egyptian pharaoh, the name itself was Greek. It was easily recognizable in the bottom, Greek portion of the stone. Noting the number of times and the placement of its occurrences, he found correlating cartouches in the upper portion. In another source, he also located the name Berenice, the wife of Ptolemy I, and another Grecian name. The symbols within the cartouches must represent letters. Young thought these letters were only used within the cartouches. He was wrong about that, and it would be left to Champollion to find the true nature of the hieroglyphic alphabet. Young made other discoveries regarding hieroglyphs, but in spite of his success in demotics, his advancement with hieroglyphs was quite limited. If he had extended his theory of demotics to the hieroglyphic symbols, he might have made great progress there as well. He believed demotic was a mixture of letters and word symbols, but he thought hieroglyphs were strictly word symbols—ideograms. He assumed the birds, turtles, triangles and all the other various shapes represented things or concepts, but not letters. It is easy to see the confusion. The demotic script does look something like a writing script made up of letters. Hieroglyphs do not. That error caused most of Young's work to be incorrect.

An example of Egyptian hieroglyphs. The oval, or cartouche, shows that the hieroglyphic symbols within formed a proper name.

Another limiting factor may have been that Young made so many discoveries in so many different fields that he did not always spend the time necessary in an area to achieve all he might have. An Italian linguist, Amedeo Peyron, spoke to this issue, both praising and admonishing Young, when he wrote to him, "You write that from time to time you will publish new material which will increase our knowledge of Egyptian matters. I am very glad to hear this, and I urge you to keep your word. For as Champollion will witness, and other friends to whom I have mentioned your name, I have always felt, and so do many others, that you are a man of rare and superhuman genius with a quick and penetrating vision, and you have the power to surpass not only myself but all the philologists of Europe so that there is universal regret that your versatility is so widely engaged in the sciences—medicine, astronomy, analysis, etc., etc., that you are unable to press on with your discoveries and bring them to that pitch of perfection which we have the right to expect from a man of your conspicuous talents; for you are constantly being drawn from one science to another, you have to turn your attention from mathematics to Greek philosophy and from that to medicine, etc."[32]

The stage was now set for Champollion, though he would have his own missteps as well. Like Young, he initially believed that hieroglyphs were not based on a phonetic alphabet. Hieroglyphs "are signs of things and not sounds," he would confidently state, through later retract.[33] He had imperfect copies of the Rosetta Stone, but he also had many other sources of ancient Egyptian writings. Though some names had already been found by Young, he would be able to independently find names within various cartouches—Cleopatra, Ptolemy, Alexander, Trajan, Caesar, and Berenice. Even though he and others were opposed to the manner of its removal from Egypt, he was able to make gains by his study of the Dendera Zodiac.

A monument known as the Bankes obelisk was another important source. William Bankes was a wealthy adventurer and collector of antiquities. He had first seen this obelisk in a sketch done by Vivant Denon, lying in its original location on the island of Philae, half-buried in the sand. In 1821, Bankes went to Egypt, loaded the obelisk on a ship, and brought it back to England. He set it up on the grounds of his mansion, where it still stands today. It contained both a hieroglyphic message and its translation in Greek and has been called a second Rosetta Stone. It contained the names of Ptolemy VIII and his wife Cleopatra. (This was Cleopatra III; the most famous person with that name was Cleopatra VII.) The name Cleopatra unlocked more letters, adding to his growing collection in the letters of an Egyptian alphabet. Like a crossword puzzle in which filling in blanks often leads to being able to fill in more, every new name and new letter was a possible clue to more discoveries.

There was another reason to believe that hieroglyphs were not entirely symbolic. The Greek portion of the Rosetta Stone contained approximately 500 words. Yet there were almost 1,500 symbols in the hieroglyphic section.[34] Champollion felt that if each hieroglyph represented a word, there should be approximately the same number of words in each. Therefore, some of those hieroglyphic markings must be part of an alphabetic system.

He believed at least a portion of that top section of the Rosetta Stone was based on a phonetic alphabet. Alphabetic systems are generally phonetic but can range from being partially to totally phonetic. For example, the English language is partially phonetic. The letter "g" is pronounced quite differently in words such as through, rough, gym, and gill. The letter "c" had different sounds in hack, tacit, and chalk. However, to know the sounds of the letters of a totally phonetic alphabet is to know how to pronounce the words. A common example would be that of a picture of a bee followed by a picture of a leaf, which yields the sound bee-leaf and thus, spells the word "belief." While it was called an alphabet, unlike letters in a language such as English, the symbols might be thought of as phonetic signals. If the letters used in the English alphabet were used in the same manner as in ancient Egypt, the word excess would be spelled "xs" and essay would be spelled "sa." Champollion found that the pronunciation of those letters came from the initial sound of a specific hieroglyph. A symbol of a sun, if used as a letter, could represent the letter "s." A symbol of a jackal, if used as a letter, could stand for the letter "j." The alphabet was something like the military alphabet of alpha, bravo, Charlie, delta, echo, foxtrot, … used to spell words. The initial sound gives the name of the letter.

By stripping away all but the initial sounds, he could reduce to a letter what might otherwise indicate a word.[35] The writing of Coptic and hieroglyphics were far different in appearance, but the sounds were the same. Because he knew Coptic, he knew the meaning of the hieroglyphic word.

Another breakthrough came after the discovery of a cartouche that was new to him. It contained the hieroglyphs ΘⲘⲤⲤ. From Young's work, he knew the last two symbols each corresponded to the sound of the letter "s." He believed Θ symbolized the sun. In Coptic, it was pronounced "ra"—the name of the Egyptian sun god. The second symbol, which looked something like an upside-down pitchfork, he thought might be the symbol for birth, which he came to find was pronounced "ms." Putting those sounds together gave the sound of Ramses. He knew that to be the proper name of several pharaohs from Egyptian history.

The theory at that time was that there was an Egyptian alphabet, but its only use was for spelling foreign names. All other words should have their own symbol representing them. Yet, here was a name, Ramses, that was very much a native Egyptian name. Hieroglyphics were much more alphabetical than previously believed. This was a monumental discovery by Champollion.

It is not completely clear from his notes whether it was this or another event that made Champollion race the 200 yards from his house to his brother's office at the Institut de France. After running that distance, he yelled *"Je tiens l'affaire"*—I've got it. At that point, he collapsed. In subsequent retellings, he was unconscious anywhere from a few hours to almost a week. Being unconscious for several days seems unlikely, especially because the story was not related until years later. However the event transpired, Champollion believed he had stumbled onto something that would totally change the thinking on how hieroglyphs were to be understood.

He put his discoveries into a paper entitled *Lettre à Monsieur Dacier.* Bon-Joseph Dacier was the secretary of the Acadèmie des Inscriptions et Belles-Lettres, an organization dedicated to the humanities. Roughly a week after Champollion's discovery, on September 27, 1822, he made a presentation to that group. De Sacy and Jomard made presentations that same day, and by coincidence, Thomas Young was in Paris at the time and also attended the lecture. They had a cordial meeting afterward, but there would later be talk of plagiarism. Young would not join in directly but did express disappointment that Champollion did not acknowledge his substantial contribution that he felt led to Champollion's discovery.

Progress then proceeded rapidly. Soon, Champollion could read almost any of the Egyptian writings. While there would continue to be work in hieroglyphs, even to the present day, for the most part Champollion had released Egypt's secret language. He had found that hieroglyphs could be interpreted in several different ways. A hieroglyphic symbol could be an ideogram, representing words, phrases, objects, or ideas. A present-day example might be that of a light bulb drawn above a head, representing the concept of an idea. A hieroglyph could also be a pictogram, a literal representation of an object. An image of a cat is simply a cat. Finally, the symbols could also be part of a phonetic alphabet, with that alphabet spelling out its pronunciation. The hieroglyphic sound was the same as in Coptic, and Champollion knew Coptic.

Another discovery of Champollion's was a category of symbols known as determinatives. A determinative mark could be added to a word to help when more than one meaning was possible. Did the symbol Θ refer to the sun or the sun god Ra? Did a symbol represent a word or a letter? Also, since there was no spacing between spelled out words, a determinative could be used to show when a word was finished. Determinatives were not words or letters, but necessary to clarify meaning.

The Final Years

Champollion's later life continued to be remarkable. He went to Italy in June 1824 in search of more source material—a good portion of which had been taken by Napoléon, but then returned after his fall. Champollion had met with Louis XVIII who agreed to finance the trip. He visited the Vatican and other sites in Rome, as well as Turin, Naples, Florence, and even Pompeii, which contained documents that had been carbonized following the eruption of Mt. Vesuvius.[36] He was able to obtain and read from the Egyptian Book of the Dead—the first person to do so in 2,000 years. In the fall of 1825, he returned home, just before the Alps became impassable, to be reunited with his wife and daughter whom he had not seen for a year and a half.

In May of 1826, he was appointed the first curator of Egyptian Antiquities at the Louvre Museum. He made major purchases which added to the Louvre's collection, as opposed to the Napoléonic tactic of simply taking valuable items from other lands. Edme Jomard was upset that Champollion got this position over himself, and it added to his already abundant bitterness. In 1831, Champollion was appointed to the chair of Egyptian archaeology at Collège de France. It was a newly created position and the first ever of its type.

There would be another adventure that same year. This one fulfilled his life-long dream of actually going to the land of Egypt. It was a joint venture with Ippolito Rosellini, Italian Professor of Oriental Languages.

They landed in Alexandria in August 1828. The French and British had departed, Ottoman rule was restored, and the reign of the Mamelukes had come to an end. There he met with the current leader of Egypt, Muhammad Ali, who had been ruling there for over 20 years when he met with Champollion and Rosellini. Ali was in the process of making improvements to his country, but as when the Rosetta Stone was found among rubble, that often meant tearing down and repurposing ancient buildings to make way for the new. Champollion wanted access to Egypt before there was more damage to monuments from being destroyed in rebuilding programs or looted by thieves. An agreement was reached with Ali that allowed Champollion to sail the Nile River as far as the second cataract. His crew of 30 would spend over a year traveling the Nile, following much the same route as Vivant Denon three decades prior.

Champollion looked the part. Already dark-complexioned, he shaved his head; wore a turban, baggy trousers, and a full beard; and of course, spoke fluid Arabic. He was able to read messages that had not been read by anyone for centuries. Egypt's history was literally spelled out before him on the temples, pyramids, and obelisks he visited. The natives called him, "The man who could read the writing on the old stones."[37]

He was able to see Giza with its pyramids and the Sphinx, still covered in sand to its neck. He wished to have that sand removed, but he was told by the natives it would take 40 men over a week to make that happen.[38] He visited Thebes, Abu Simbel, Dendera, Luxor, Karnak, Aswan, Philae, and the Valley of the Kings. It was for him a joy and a fulfillment of a life's work.

He spent hours inside pyramids doing far more than the earlier savants, who could simply record what they saw, while he could actually read and make sense of the hieroglyphic writings. He had to take breaks to get fresh, outside air, and even with that, he fainted on several occasions. At times, he had to be carried because of his gout. At this point of his life, he was quite overweight, making his explorations even more difficult. He described one of his journeys inside a pyramid—"I undressed almost completely, down to my Arab shirt and long linen underpants, and pushed myself flat on my stomach through the small opening in the doorway that, if cleared of sand, would be at least 25 feet in height. I thought I was entering the mouth of a furnace, and, when I had slid entirely into the temple, I found myself in an atmosphere heated to 52 degrees [126° Fahrenheit]: we went through this astonishing excavation, Rosellini, Ricci, I and one of the Arabs holding a candle in his hand."[39]

They arrived back in Toulon, France on December 23, 1829, having spent 15 months away. Before he could travel further, though, his company had to spend a month in quarantine because the plague was still in full force in Egypt. He was able to bring a hundred artifacts back with him, and they would find their places in the Louvre.

He would live another two and a half years, dying at age 41, on March 4, 1832. The cause of his death was never clearly established. Perhaps it was some combination of his history of health problems, his weight, a driven personality, the strain of the trip, or an illness he contracted in Egypt. His family was with him when he passed. They related his final words: "And now the afterlife, on to Egypt, on to Thebes."[40]

His brother would live many more years, dying at age 89 in 1867. In the following years, Jacques-Joseph would collect, edit, and publish much of his brother's work in the same manner that Marie Lavoisier had published that of her husband, Antoine.[41] Jacques-Joseph lost his professorship in the French Revolution of 1848 but was later appointed Keeper of the Library of the Palace of Fontainebleau. He would never fulfill his dream of going to Egypt.

Jean-François Champollion was buried in the Père Lachaise Cemetery in Paris, his grave marked by a stone in the shape of an Egyptian obelisk.

VIII

The Visionaries

The discoveries of penicillin, the planet Neptune, or the printing press happened relatively quickly and by the work of one or two individuals. Some discoveries, though, have many steps along the way, with many individuals making contributions during that time. One person's work often leads to another's. The latter person may come to be more well known, yet his or her discovery may well not have taken place if not for the work of those that came before. French men such as Lamarck, Cugnot, Niépce, Chappe, and Jacquard are not well known, but laid the groundwork for others.

Evolution

In the preface of her novel *Frankenstein*, Mary Shelley wrote, "The event on which this fiction is founded has been supposed by Dr. Darwin, and some of the physiological writers of Germany, as not of impossible occurrence." Shelley was not speaking of the famed Charles Darwin, but his grandfather, Dr. Erasmus Darwin. Erasmus was not only a physician, but also one of England's most famous poets, and one who had things to say regarding evolution.

He was also a Unitarian, having dined with Unitarian preacher, chemist, and foil to Antoine Lavoisier, Joseph Priestley, at Priestley's Birmingham home.[1] His famous grandson, Charles, would be raised in the Unitarian Church, though his religious beliefs would later go astray.[2] Erasmus wrote *Zoonomia*, a large, influential poem that said, in part, "...would it be too bold to imagine, that all warm-blooded animals have arisen from one living filament...."[3] He wrote it in poetic language rather than in prose, thinking that might make his statements a bit less controversial. It did not. Critics used words like "monstrosity," "shocking," "atheism," and "degradation" in commenting on the work.[4]

Charles Darwin has been credited as the first to consider the concept of evolution. However, he was not the first, even within his own family. It was already known that animals change, and it had even been theorized that there could be change from species to species. He was, however, the first to have an explanation of exactly how that change might take place. He called his theory natural selection. A breakthrough moment for Darwin took place a half world away from his England home.

Thoughts on evolution, called "descent" by Darwin and often called "transmutation" by others of his time, is almost as old as mankind's history. Greek and Roman philosophers wondered where humans came from and if there was a progression from one species to another. Anaximander, in the sixth century BCE, thought life originated in the sea and mankind is a descendent of sea creatures.[5] While that sounds Darwinian, he thought this process came about when humans burst forth from the inside of fish. Philosopher Ibn Khaldun thought life progressed gradually from minerals to plants to animals.[6] Many thought that the animal world began when creatures simply emerged from the mud.

Though Darwin is by far the most famous name in evolutionary thought, his writings would not have been possible without the work of three French scientists—Cuvier, Geoffroy, and Lamarck. All three were well-known naturalists and were associated with perhaps the finest botanical garden in the world, the Jardin des Plantes (Garden of Plants). It received that name in 1793, when the National Convention changed its name from the Jardin des Roi, since "Roi," being the French word for king, would obviously no longer be acceptable. The garden would house the Museum of Natural History, created in the summer of 1793, the same year as the Louvre Museum.

George Cuvier was born in Montbéliard in 1769. His hometown was technically just outside the French border, but during the wars of the French Revolution, it became a part of France and remains so today. His family was part of the bourgeoisie, his father being a lifelong military officer. George excelled in school. He read and studied the immense *Histoire Naturelle*, written by Georges-Louis Leclerc, Comte de Buffon. Buffon was born in 1707 into a wealthy, noble family. He was appointed director of the then-named Jardin du Roi, which he improved and enlarged. His *Histoire Naturelle* was a summation of what was known of science up to that time, containing 36 volumes that he wrote, with more being written by others after his death. It was translated into several languages and valued throughout Europe. Buffon died in 1788, a year before the storming of the Bastille, thus avoiding the years of revolution. However, he was, in a sense, affected even after his passing. His tomb was razed by rioters seeking lead for bullets and his son would be sent to the guillotine.

Like many young French men of that time planning their futures,

Cuvier intended to enter the clergy, though he later changed his mind. For him, and many others in France, there was a decreased emphasis regarding religion. Part of that was due to enlightenment thought, emphasizing man and deemphasizing God. Lagrange purportedly said, when asked why his writings did not mention God, "I have no need of that hypothesis." The church was not done away with but had much less visibility in society. Priests who did not sign the oath of allegiance to the state laid low, left the country, or were arrested, while schools run by priests, of which there were many, were closed and their land taken. Cuvier ended up attending a school in Germany. After graduating, he lived with a noble family for several years, serving as a tutor to the Comte d'Héricy. When revolutionary activity began to heat up, the d'Héricy family opted to remain in France, but moved to coastal Normandy, further away from turmoil. This turned out to be an excellent opportunity for Cuvier, expanding his knowledge by allowing him to study marine life on a daily basis.

He wrote papers on his findings and sent them to naturalist Geoffrey Saint-Hilaire, a Paris professor who was recently a part of the French expedition to Egypt. Cuvier's papers were of such quality that they led to his gaining a position as a natural history professor in Paris. It was the beginning of a working relationship between the two, even resulting in their coauthoring many papers together. Cuvier found an abundance of specimens to study in Paris, thanks in part to the many artifacts Geoffrey Saint-Hilaire had brought back from Egypt.

Cuvier developed a classification system, dividing the animal kingdom into four groups, which he called vertebrates, articulates, mollusks, and radiates.[7] He saw this as an improvement on a classification system that had been developed by Swedish naturalist Carl Linnaeus, considered the founder of modern taxonomy, who developed his system in 1735.[8] His system used two Latin names to refer to species, such as Tyrannosaurus rex or Homo sapiens. Linnaeus did not believe in evolution. He believed species were fixed. What is the point of classifying species if they could change from one to another?

Cuvier spent much time finding and examining fossils. He is considered the father of paleontology, the study of the past through plant and animal fossils.[9] He took bones he found and reconstructed skeletons of animals no one living had ever seen, nor had ever been named. He declared that he found fossils, such as the mastodon (named by him), unlike anything living on earth at the time. That species could become extinct was a new, controversial idea. Benjamin Franklin spoke for many when he said, "No species [of animals] or genus of plants was ever lost or ever will be while the world continues."[10]

Cuvier stated that information from fossils could lead to the geologic dating of earth's history. He studied fossils of the Paris Basin with Alexandre Brongniart. (Brongniart would go on to name a specific time period, the Jurassic Era, while exploring in the Jura mountains.) As the two dug further down, they found different layers of fossils. Interestingly, the deeper they dug, the more dissimilar the fossils were to current life forms. They felt this showed a progressive change in animal life through time, serving as a calendar of earth's lifespan

While Cuvier did recognize that changes took place, he discounted the concept of evolution. He said species that became extinct did so due to a series of cataclysmic events—a theory he labeled catastrophism. Events like ice ages and floods, the flood recorded in the book of Genesis being the last of these, were what caused entire species to become extinct. If other species of animals did take their place, it was by a process other than that of evolution. His disbelief in evolutionary change would result in a falling out with both Geoffrey and Lamarck. Cuvier cited mummified animals that had been brought back from Egypt. They looked to be unchanged as compared to present day versions, even though thousands of years had passed. An evolutionist, such as Charles Darwin, would rebut Cuvier by saying that even thousands of years is not nearly enough time to see evolutionary change take place. Cuvier felt that the species currently alive were well structured and well fit to their environment. For him, that was evidence that species had no need for change. Again, one such as Darwin would argue that species, including humans, had changed and further change was not out of the question.

Étienne Geoffrey Saint-Hilaire, commonly called Geoffrey, worked with and mentored Georges Cuvier, but the two would have sharp disagreements later regarding evolutionary theory. Geoffrey was born just south of Paris, in Étampes, in 1772. He became a naturalist, studying at the Collège du Cardinal in Paris. During the revolution, Hilaire risked his life to save fellow students and teachers, even organizing a nighttime escape attempt from their prison cells.[11]

In 1798, Geoffrey was one of the savants who accompanied Bonaparte to Egypt. (Cuvier had been invited to go to Egypt but chose not to.) Geoffrey would continue to be a supporter of him even after the inglorious end of the expedition, though many other savants left stranded in Egypt did not feel the same way. Geoffrey was the key person in getting the huge number of specimens the savants had collected, both living and dead, back to France. By 1801, the British and Ottoman forces had clearly defeated the French forces. The British position was that they were the victors and deserved the spoils. They certainly were going to keep the Rosetta Stone, already seen as a valuable artifact, but other items were up for negotiation. Geoffrey valued his collection and threatened to destroy them all rather than turn them over to the British. The British relented and allowed the French, and a good portion of their artifacts, safe passage back to France. French naturalists were enriched by the specimens Geoffrey brought back with him. In future years, Napoléon would send him in search of other specimens from other areas of the world.

At the height of the revolution, Geoffrey moved back to his parents' house where he had something at least resembling a nervous breakdown. He was bedridden for months. Geoffrey had already experienced difficulties with his health while in Egypt. There, he became ill and lost a good deal of weight, to the point that fellow savants became quite concerned for his life. Concern was not only for his physical well-being, but also for his agitated mind.[12] Also, like so many others that went to Egypt, he had periods of blindness due to repeated bouts of ophthalmia. He lived another forty years after his return to France, so there may not be a connection to his difficulties in Egypt, but for the final years of his life he was in poor health and then totally blinded by a stroke.

Cuvier and Geoffrey grew apart. Cuvier was very religious and saw in nature the skill of the Creator, while Geoffrey became increasingly agnostic in his beliefs. Cuvier did not believe in evolution, while Geoffrey did. They disagreed on what is called comparative anatomy, comparing body structures of different species, an area Cuvier is considered to have founded.[13] They argued about the relationship between various animals' arms, wings, fins, or flippers. Are they different, unrelated appendages or linked in an evolutionary past? Cuvier said that forms were established at their creation while Geoffrey said that a species' anatomy is shaped over time, depending on its environment. Cuvier: Giraffes have always been tall. That is why they eat from the tops of trees. Geoffrey: Giraffes sought the leaves at the tops of trees and evolved to be taller. They displayed their contrasting views in a series of eight well-attended public debates.

Jean-Baptiste Lamarck was another major influence on the thinking of Charles Darwin. Lamarckism was a theory that animals acquired traits during their lifetime that could be passed on to future generations. To again take the case of giraffes, they are tall not because they have always been so (according to Cuvier), or because the tallest survived and passed their genes on (Darwin), but because the process of reaching to the tops of trees naturally caused their necks to lengthen. Lamarckism claims that acquired physical characteristics, such as growing a longer neck or becoming stronger through exercise, can be passed to offspring. Though differing from Darwin's concept of natural selection, they both shared the concept that species could undergo change.

Jean-Baptiste Lamarck, born in 1744 in Bazentin, France, was the youngest of 11 children. At the turn of the century, he was in his fifties, something of a senior statesman of the trio, while Cuvier and Geoffrey were in their thirties. He had poor eyesight throughout his lifetime and was totally blind his last ten years. He died in 1829, and his finances were such that his family had to borrow money to pay for his funeral.[14]

He had intended to become a priest, but the death of his father and the expulsion of Jesuits changed those plans. He began a military career, fought in Europe in the Seven Years' War, and was recognized for his bravery. He became a professor of natural history and was a protégé of Comte de Buffon. Lamarck would coin the terms "biology" and "invertebrates."[15] Seemingly the most scientifically well-rounded of these men, he did work in a number of areas including physics and chemistry.

Charles Darwin was born in Shrewsbury, England, February 12, 1809, the same day, across the Atlantic, that Abraham Lincoln was born. Lamarck was poor, but Darwin was extremely well-off financially, thanks to the family he was born into and due to his own wise investments as an adult. Charles' mother died when he was eight years old. His controlling father decided Charles would follow in his own footsteps and sent him to Edinburgh to become a doctor. Though bored with the medical curriculum, the knowledge acquired would help him in his later work. With things not working out in Edinburgh, his father sent him to Cambridge to be a clergyman.

Having taken an interest in the study of plants and animals, he served as the ship's botanist on the *HMS Beagle*, with a crew sent on an around-the-world journey, primarily to map the coastline of South America. It was on this trip, at the age of 22, that he found himself on the Galapagos Islands. While he encountered many animals new to him, the most important would be the small finches on the islands. He found 13 different types of finches—different types on different islands. Their beaks seemed to be of a size and shape that was best suited to the kind of food available on each island. His theory of natural selection sought to explain why. If a finch had a certain kind of beak that worked well for a certain kind of food, the possessor of that beak would be more likely to survive and have its DNA passed on to its offspring. On the other hand, finches with inferior beaks would not be as healthy, likely having fewer children, or simply dying young.

Darwin found merit in Lamarck's ideas. His own theory did not come to him on the Galapagos Islands, and, as in the case of evolutionary change, it would be many years in the making. In fact, it would be 20 years until he published *On the Origin of Species by Means of Natural Selection*. It was another 12 years later that he published the equally controversial *Descent of Man*.

Charles Darwin was not the first to consider the concept of evolution, but his theory of natural selection supplanted Lamarckism and has held sway to current times. However, it has continued to be controversial and final chapters have not necessarily been written. In the 1970s, paleontologist Stephen Jay Gould stated that the evolutionary process was not nearly as drawn out as Darwin thought. He felt fossil evidence showed there are long periods of time of little or no evolutionary activity punctuated with relatively sudden bursts of change.[16] After the popularization of Darwin's theory, Lamarckism faded from view, even being held in contempt as something akin to a flat earth theory. However, Lamarckism may play a role after all. For example, many evolutionist scientists now believe that genes may be turned on or off due to factors in the environment, and that process can then be passed on to future generations—a field now known as epigenetics.[17]

The Automobile

The automobile is one invention that went through several stages of development. The vehicle built by Nicolas Cugnot did not go fast or far, but he is credited with inventing the first self-propelled land vehicle.

Nicolas-Joseph Cugnot was born in the city of Void, in northeastern France, in 1725. Captain Cugnot was an artillery officer and engineer. His military service ended with the conclusion of the Seven Years' War at which point he became a teacher at the Paris Arsenal, where he also wrote books on military fortifications. His background prepared him well for his next endeavor—increasing the speed at which troops can be moved. A large part of General Bonaparte's success in battle was the speed with which his armies moved, being able to outflank an opponent or to get to a desired position first.

The armies of the day were made up of cavalry (mounted troops), infantry (foot soldiers), and artillery (heavy guns), whose speed generally decreased in that order. The overall speed of an army depended on how quickly it could move its heaviest weapons. Heavy artillery often had wheels, so it could be pulled by manpower or horsepower. It could also be loaded onto drays—strong carts, typically pulled by horses, built for the purpose of hauling heavy loads.

In 1769, Cugnot built a self-powered vehicle; then built a larger one the following year. Powered by steam, they were heavy, strong, and capable of hauling objects such as cannons. He designed his steam engine at about the same time, and independently of the English inventor James Watt. Cugnot's vehicle had two wheels in back and a smaller one in front with a large boiler located over the front wheel. There was fire at the bottom of the boiler which heated the water above it, turning it to steam and powering its pistons. There were two vertical pistons, one on each side of the front wheel. The up and down motion of the pistons was converted to a rotary motion capable of turning the wheel. Though cumbersome and only a first step, Nicolas Cugnot had created an automobile.

Cugnot's vehicle was not fast. He was hoping for a speed of something in the five to ten miles per hour range. It went just over two miles per hour; not much more than a very leisurely walking pace.[18] It could run for about 20 minutes, then would have to take on more water. It wasn't very maneuverable and, having only three wheels, it potentially would have stability issues in rough terrain.

According to reports, in 1771, Cugnot's invention knocked down a wall in Paris for which he was arrested. If so, Cugnot, in addition to creating the first car, was in the first car accident, and was the first to be arrested for reckless driving.[19] However, this incident was first reported as part of his obituary when he passed away thirty years later, so is of questionable validity.

Cugnot had received some initial government money, but when France's foreign minister, who had been very supportive of his work, was replaced, that money went away,[20] though the king rewarded his work with a pension. That royal pension was taken away at the onset of the revolution, and he was exiled to Belgium where he lived in poverty. When Napoléon Bonaparte came to power, he invited Cugnot back to France and restored his pension, which he was able to enjoy for a short time before his death in 1804. Cugnot's car can be found today in the Musée des Arts et Métiers (Arts and Crafts Museum) in Paris along with many other displays, such as Antoine Lavoisier's laboratory and a Jacquard loom.

The work continued, with several Frenchmen making significant contributions. Jouffroy d'Abbans built the first steamboat. Like Cugnot, he was an army officer, though he was incarcerated for a time for dueling with a superior officer.[21] His initial steamboat attempt was a failure, but on July 15, 1783 (just one month after the Montgolfier brothers' first public demonstration of flight), he made a public showing of his boat. It was 140 feet long with paddle wheels on each side and a steam engine between them.[22]

As in the case of the Montgolfier brothers, the Academy of Sciences wanted to see it demonstrated to them before giving their stamp of approval. However, d'Abbans simply did not have the personal funds to transport his invention to Paris. He sought a remedy through the courts which is where things stood when the revolution came. Later, because he was a supporter of the king, he emigrated and later fought against the French. After the turmoil had passed, he returned to live the rest of his life in France, passing away in Paris in 1832. Robert Fulton, who was influenced by d'Abbans' work, would later build the first commercially successful steamboat.[23] While d'Abbans' efforts came to an abrupt halt, Fulton, though an American, made his first public demonstration on the Seine River in France.

Phillipe Lebon improved on Cugnot's automobile design. He obtained a patent on an engine fueled by coal gas with an electric ignition, though it was never actually built. He also did important work in gas lighting and heating. In 1799, his "thermolampe" burned gas from distilled wood and led to the lighting of streets in the nineteenth century.[24] He was invited to help plan the festivities surrounding Napoléon's coronation in 1804. Tragically, Lebon was killed on the day of the ceremony, apparently by thieves, although the circumstances of his demise remain unclear.

In 1807, Isaac de Rivaz created a gas-powered vehicle that used hydrogen gas as a fuel and was ignited by a spark from a Volta cell. That, in itself, was a major advancement, but the following year he placed it inside the vehicle, making it the first internal combustion powered automobile. That same year, Nicéphore Niépce, who would later make history in the field of photography, was doing much the same as Rivaz. He created the first internal combustion powered boat. Though his was a boat and Rivaz's was a car, both are typically given credit as inventors of the internal combustion engine.

Interchangeable Parts

Honoré Blanc was born in Avignon, France in 1736, and was trained early in life to be a gunsmith. There was plenty of work for those in his profession. Manufactured rifles were unique. Blacksmiths forged each one, and the parts from one would rarely be a perfect fit in another. If a firing mechanism broke, it required sending the gun to a skilled gunsmith to make the repair.

The firing mechanism, or lock, was the part that was likely to malfunction. The proverbial "lock, stock, and barrel" were the three parts of a rifle. The stock, the wooden portion held against the shooter's shoulder, and the barrel, the metal tube through which the bullet traveled, rarely broke. The lock, though, was a complicated mechanism with several breakable parts. The cost and time of sending the firearm off to a skilled gunsmith to fix was inconvenient, but especially so if this took place on the battlefield.

Charles Dickens knew about this difficulty when he wrote of a military officer faced with a broken lock and seeking the aid of a black-smith named Joe in his novel *Great Expectations*—"You see, blacksmith," said the sergeant, who had by this time picked out Joe with his eye, "we have had an accident with these, and I find the lock of one of 'em goes wrong, and the coupling don't act pretty. As they are wanted for immediate service, will you throw your eye over them?"

Joe threw his eye over them, and pronounced that the job would necessitate the lighting of his forge fire, and would take nearer two hours than one. "Will it? Then will you set about it at once, blacksmith?" said the off-hand sergeant, "as it's on his Majesty's service. And if my men can bear a hand anywhere, they'll make themselves useful."

In 1791, Blanc set up a demonstration showing how he had constructed the parts of the lock with such precision that he could replace a broken piece quickly and efficiently. He separated the pieces of 50 locks and an assortment of screws, plates, and springs, into bins. Individuals gathered there were shown how he could take random pieces out of those bins and assemble them into a working lock.[25] This group, and later the French Academy of Sciences, were impressed. However, the newly established Legislative Assembly was not as impressed, and chose not to fund his work.

The United States ambassador to France, Thomas Jefferson, was quite enthralled, though. He wrote to John Jay, the U.S. foreign secretary, "An improvement is made here in the construction of the musket which it may be interesting to Congress to know. It consists of making every part of them so exactly alike that what belongs to anyone may be used for every one musket in the magazine. I put several together myself taking pieces at hazard as they came to hand, and they fitted in the most perfect manner. The advantages of this, when arms need repair, are evident."[26]

Jefferson sought to bring Blanc to the United States, and though he chose not to go, Jefferson was able to convince President Washington of the viability of Blanc's work. Ironically, the U.S. government was interested in improving its fighting capability as warfare with France seemed a distinct possibility at the time. Eli Whitney, the recent inventor of the cotton gin, obtained a contract to construct 10,000 muskets. To do so, he used information gleaned from Blanc's writings.

Honoré Blanc had pioneered the use of interchangeable parts, but gunsmiths of the time saw how damaging this could be to their own businesses. During those financially troubled times, competitors who realized Blanc's work could be industry-changing destroyed his work area. He would die in debt. However, advancements with interchangeable parts continued. The concept would fuel the industrial revolution and the efforts of future entrepreneurs.

Food Preservation

Perhaps not as sensational as creating the automobile or the photograph, a Frenchman revolutionized food preservation. While not dramatic, it would be difficult to overestimate the importance of what he did and how it affected the world.

Nicolas Appert was born in Châlons-sur-Marne in 1749. His family owned an inn and much of his early culinary preparation undoubtedly happened there. He apprenticed in the Palais Royal Hotel in Châlons, then worked in the kitchen of the Duke and Duchess of Deux-Ponts. He went into business on his own and became a noted chef and confectioner in Paris for the next ten years. His loyalties came under suspicion during The Terror, and he was arrested in April 1794.[27] Luckily, he managed to avoid the guillotine and was released. The following year, Appert began experimenting with ways to preserve food.

People had been trying to preserve food from earliest times, but with limited success. Salting meats and fish helped preserve them to a degree. Techniques such as keeping foods cold, smoking, or fermentation also extended the shelf life of foods. Prior to 1800, it was a race against the clock to get food from the farm or the field, to storage, then to the household table without spoilage setting in. A great concern to military men such as Napoléon Bonaparte was how to feed an army on the march far from home. Campaigns generally took place during the summer and fall months simply so armies could live off the land as they traveled. The advancements in food preservation would change the calendar of military campaigns. In addition to the issue of armies on the march was the problem of how to feed a navy at sea. Naval tours were often limited by the amount of food that could be taken on board and how long that food would stay edible. Appert would later write, "The chief importance of this process lies in its subservience to the wants of civil and military hospitals and particularly of the navy."[28]

To assist its military, the French government offered a reward of 12,000 francs for what it called a cheap, effective method for preserving large amounts of food. Many discoveries during this time came about due to the government's, then Napoléon Bonaparte's, wise use of prize money to motivate potential inventors. Appert sought this prize money and spent 15 years of experimentation before finding a process that worked.

The method he found used glass bottles because glass was "the matter most impenetrable by air."[29] He had special wide-mouth bottles made, placed various foods inside, corked the bottles, then reinforced the seal with wax and wire. The bottles were then placed in a kettle of boiling water. The method became known as Appertizing. "This method," he wrote, "is not a vain theory. It is the fruit of reflection, investigation, long attention, and numerous experiments, the results of which, for more than ten years, having been so surprising, that notwithstanding the proof acquired by repeated practice, that provisions may be preserved two, three and six years. There are many who still refuse to credit the fact."[30] The British quickly adopted his techniques but used metal cans instead because they were safer to transport than breakable bottles. It was many years before an effective method was found for opening those cans, with much frustration and even injuries taking place in the meantime. The can opener would not be invented for another 50 years.

Appert wrote a book titled *The Art of Preserving All Kinds of Animal and Vegetable Substances for Several Years.* It was a collection of instructions for how to preserve various kinds of foods, with chapters such as Chapter 8—Gravy, Chapter 11—New Laid Eggs, Chapter 12—Milk, and Chapter 35—Strawberries. He won the government-sponsored contest and used that money to open his own bottling factory—the first of its kind. His House of Appert operated in France for more than 100 years. It was not only a commercial site to preserve food, but also a location which Appert used to continue his research. He improved his process, experimenting with using cans rather than bottles, and made other discoveries, which included creating the bouillon cube.[31] His factory had a difficult beginning. It was destroyed as forces of the Sixth Coalition chased Napoléon Bonaparte's troops back from Russia. The French government did help him rebuild, however, and the House of Appert operated until 1933.

Nicolas Appert changed the world, and doubtlessly saved countless lives. His process worked, but he never knew exactly why it worked. It would be another 50 years until fellow countryman Louis Pasteur explained that the method worked because the use of heat killed off the microorganisms that caused food to spoil and the sealing kept others out.[32]

The Optical Telegraph

"Two inventions seem to have marked the 18th century; both belong to the French nation: the balloon and the telegraph ..."—French politician, Joseph Lakanal.[33]

In 490 BC, the Battle of Marathon was a Greek victory over the invading Persians. It was said that Pheidippides ran roughly 25 miles carrying the news back to the city of Athens. That was telecommunication at the time. Some progress was made in the centuries since. People used drums, smoke, or messenger pigeons to send messages, but an effective system did not come about until the work of Claude Chappe.

Chappe was born into a wealthy family in Brûlon, France in 1769. He studied to enter the church, but that was cut short by the French Revolution. He came home, and with his four brothers, began to work on what they hoped would be a method of improving how messages could be sent.

After some initial, less than successful, trials, they settled on sending messages through a series of towers, roughly ten miles apart. Each tower would have a crossbar that could be rotated to one of four positions—horizontal, vertical, or 45 degrees to the left or right. Two shorter bars were attached to the ends of that crossbar. Each of them could be in one of seven positions, 45 degrees apart (all except the case where it would lie on top of the main crossbar). As a whole, the towers looked like a human figure with outstretched arms which could freely rotate its forearms at the elbows. With the four positions of the main crossbar and seven on the two attached pieces, there were $4 \times 7 \times 7$, or 192, possible combinations. Letters, numbers, or common words could be assigned to each of those positions, although ultimately only 92 positions were used in the Chappe system.[34] The tower at each station contained two telescopes to see the arm positions at the adjacent stations, and a code book containing the list of translations. Claude coined the term "semaphore," Greek for "bearing a sign."

Brother Ignace Chappe was a member of the Legislative Assembly and helped push through the proposed system. Final approval came from the assembly's successor, the National Convention. It was adopted as a public utility and used primarily for government and military concerns, though it

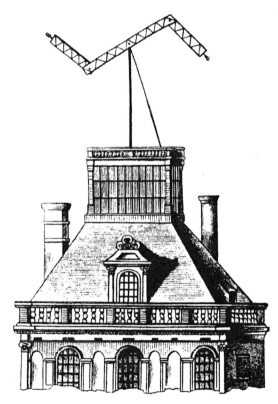

A depiction of one of Claude Chappe's optical telegraph stations. The stations sat atop buildings or towers so they could be viewed from a distance.

was also used to communicate what was apparently information of equal importance—the latest winning lottery numbers.

The first major route stretched from Paris to Lille, almost 150 miles apart, with 16 stations between the two cities. Lille, at the time, was on the war front. In August, a message was sent from there letting Paris know of the French victory over Austrian forces at Condé-sur-l'Escout. The delivery of the message took less than an hour. Another line was built stretching from Paris to Toulon on the Mediterranean coast. It took 116 stations to span the 500 miles, and messages could be delivered in 20 minutes.[35] Many other lines were constructed, both in France and through its newly conquered territories. The system would be adopted by other countries.

Napoléon made use of the system both as a leader of armies, and later as the leader of the country. His troops held a communication advantage over the various coalitions, because not only did the coalition armies speak different languages but they also did not possess the ability to share information as quickly as the French. He used it to solidify support during the Coup of 18 Brumaire. During The Hundred Days, this optical system provided warning to the new king, Louis XVIII, and gave him time to escape Paris before Napoléon's entry. Unfortunately, it was also used to notify Paris of Napoléon's victory at Waterloo, which then had to be corrected two days later.[36]

The optical telegraph became an electric telegraph system a few decades later, after its creation by American inventor Samuel F. Morse. The other four Chappe brothers stayed involved, administering their system until they were ousted in the Revolution of 1830. Claude had issues with depression throughout his life along with various other health concerns. After the implementation of his telegraph, there were charges of him stealing ideas from others. In the end, Claude Chappe committed suicide, although it is not exactly clear why. Whatever the reason, Claude Chappe took his own life in Paris on January 23, 1805.[37]

Computer Programming

Weaving is the interlacing of threads to make fabric. It has been used since ancient times to make items such as blankets, clothing, and tapestries, being done either by hand or by machine. If by machine, a device known as a loom is used, which was first used in China more than 2,000 years ago. When using a loom, a set of parallel threads are held in place, then interlaced at right angles by other threads in an over and under pattern. If there are no changes, the process is fairly straightforward.

However, if the fabric is to have some type of pattern, the process is much more difficult and much more time consuming.

To weave a pattern, different colored cords must lie on top of the weave. This was accomplished by metal rods with hooks on the ends, pulling up certain threads, allowing one to go over or under another. Each row might require that a different set of hooks be lifted, allowing different threads to pass over or under. At least two people were needed to make this work. One person could operate the loom and another, often a child, could lift the appropriate threads for each row.

Joseph Jacquard developed a machine that instructed the loom which threads to lift. That set of directions could be planned, or programmed, before any actual weaving took place. It revolutionized the weaving industry, but perhaps as importantly, also became the first instance of programming a machine, and was a direct ancestor of computer programming.

Joseph-Marie Jacquard was born in 1752 in the city of Lyon, which for the previous 300 years had been the center of the silk weaving industry. Joseph's father was a weaver, and much of Jacquard's training surely would have come from working in his father's shop as a boy. Joseph's mother died when he was ten years old, and only one of his eight siblings would live to adulthood. When Joseph was 20, his father died, and he inherited the family business.

During the French Revolution, the rioters went after those who supported the king, but the king's supporters often battled back. In 1793, Jacquard and his son Jean-Marie fought on the side of counter-revolutionaries seeking to defend Lyon.[38] The city was overrun by the revolutionaries, but he and his son were able to escape capture. Years later, they would both fight in the French army against foreign coalition forces. Jean-Marie would be killed during the fighting and Joseph injured.[39]

Back home, Jacquard returned to his weaving business and created a device that would make his life, and many weavers' lives, much easier. To create a fabric containing some type of pattern was a long, tedious process—progressing at a rate of roughly one inch of fabric per day.[40] In the middle of the eighteenth century, Jacques de Vaucanson developed a way to communicate directions to a loom—specifically to the rods, telling them when to lift a thread and when to leave it in place. He was an inventor that had an interest in creating mechanical objects, even, oddly, building a mechanical duck. His weaving apparatus was a series of holes on a paper that was formed into a cylinder, giving directions to the loom as it rotated. The paper would sometimes tear and there was a limit to how many lines of instruction could be put on the paper. His method was not efficient and did not catch on.

Jacquard's creation was similar in some respects but made important improvements. He put the punched holes onto cards, each line corresponding to a line of fabric. The cards were tied one to the next, so that instructions could be given indefinitely simply by attaching more cards, often using over 2,000 cards.[41] The looms themselves remained as they were, but when his device was attached, the entire contraption was known as a Jacquard loom.

The success of this loom was, in part, due to the interest in the project by an important man in the French government. Lazare Carnot would encourage Jacquard's efforts and persuade the French government to recognize his work as important and worthy of their support. Carnot was a member of an accomplished family—his son would be a pioneer in the field of thermodynamics and his grandson, the president of France. Lazare, as a young man, served as military engineer in the French army and became an accomplished mathematician. Later, he was a politician, and much of the success of the French armies would be attributed to his organizational skills. He became a part of almost every version of the various French legislative bodies. He was also in the initial group that formed the five-man Directory. Carnot was a part of the Committee of Public Safety, though he often disagreed with Robespierre and would contribute to his ultimate downfall. A supporter of Napoléon Bonaparte, Carnot was exiled after his defeat at Waterloo.

In 1801, Jacquard displayed his invention to the world at an exhibition of French industry. These exhibitions were the creation of Emperor Napoléon, and, prompted by Carnot, Napoléon personally visited Jacquard to see his machine. He declared it public property

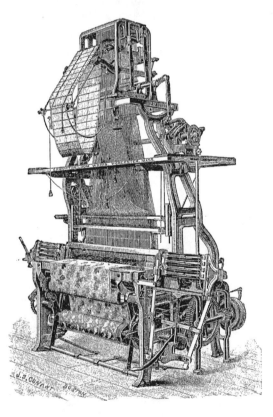

A depiction of a Jacquard loom that contains a queue of punched cards ready to give instructions.

and gave Jacquard a lifetime pension, taking him from a life of poverty to one of prosperity.[42]

The Englishman, Charles Babbage, is considered the father of the computer, developing calculating machines he called the Difference Engine and the Analytical Engine. Though Babbage was not able to complete either during his lifetime, a working Difference Engine was built in the 1820s. He saw the value of Jacquard's punched card system and planned to use it to program his computers (though the word "programming" was still in the future and "computers" applied to people that did computing). Babbage said, "You are aware that the system of cards which Jacard [*sic*] invented are the means by which we can communicate to a very ordinary loom, orders to weave any pattern that may be desired. Availing myself of the same beautiful invention, I have by similar means communicated to my Calculating Engine orders to calculate any formula however complicated."[43]

When in the poem *Childe Harold's Pilgrimage* the poet Lord Byron wrote, "Is thy face like thy mother's, my fair child, Ada! Sole daughter of my house and heart?," he was writing about his daughter, Ada Lovelace. Ada is known as the first computer programmer and was to program Babbage's new machines. She said, "The distinctive characteristic of the Analytical Engine … is the introduction into it of the principal which Jacquard devised for regulating, by means of punched cards, the most complicated patterns in the fabrication of brocaded stuffs.... We may say most aptly that the Analytical Engine weaves algebraical patterns just as the Jacquard-loom weaves flowers and leaves."[44]

Jacquard's innovation saw application far beyond that of weaving. It was used to give instructions to such devices as machine tools and player pianos, and to tabulate votes during elections. Herman Hollerith used punched cards to conduct the 1890 United States Census. The previous censuses were pencil and paper affairs, with the statistics taking longer to compile each time. It was feared that soon the results would not be available before it was time for the next decade's census. Using Jacquard's creation, Hollerith was able to collect more data and to process it in a shorter amount of time.[45] As late as the 1970s, punched cards drove the computer industry, being the way instructions and data were given to computers.

Until the advent of radio, the player piano was the main source of entertainment in many homes. It worked on the same principle as the Jacquard loom. However, instead of giving instructions to hooks and thread, the medium was air. Allowing air through holes on a piano roll dictated what notes were played on the piano.

In the twentieth century, Hedwig Kiesler was a young actress who had a gift for invention. George Anthiel was a composer who was familiar with the technology of player pianos and how they could be synchronized to play together. They combined their talents to create a system to assist Allied torpedoes during World War II. Radio-controlled torpedoes could be jammed by the enemy, sending them off course. This new technology, called radio hopping, could prevent this jamming by a rapid switching of radio frequencies.[46] The switching was coded in gaps on paper much as was done with player pianos and Jacquard looms.[47]

Hedwig changed her name to become the Hollywood actress Hedy Lamarr. Her work would become the basis for Bluetooth, GPS, and Wi-Fi technology.[48]

Photography

Whether to spread a message or as a work of art, the quest to capture an image goes back to earliest times. From drawing on cave walls, to tracing silhouettes, to painting, people have sought to create images representing their world. Pinhole cameras had no lens but admitted light through a tiny opening on one side of a box, leaving an inverted copy on the opposite side. However, that image would disappear when the light did. The search to find a photochemical method gained ground as alchemists, such as Carl Schleele, the discoverer of oxygen, found substances that were sensitive to light.

Nicéphore Niépce developed a light sensitive plate that could capture an image and hold it indefinitely. He made the first permanent photographic image and is credited with creating the field of photography.[49] That first picture was made around 1826, but it is clear from his correspondence that he had been working in this area for at least a decade.[50]

That first photograph still exists. It is entitled *View from the Window at Le Gras*. To capture the image, he used a six by eight–inch pewter plate and coated it with a substance known as Bitumen of Judea. That this important event took place was not well known at the time and even the location of the photograph was unknown for much of the twentieth century. It was found in 1952 by Helmut Gernsheim, a Jewish photographer who had escaped Nazi Germany in the 1930s. The photo now exists as part of the Gernsheim Collection at the University of Texas.

Niépce was born, lived almost all his life, and died in Chalon-Sur-Saône in central France. He was born in 1765 into a wealthy family, which is usually advantageous, but during the French Revolution it often was not. He laid low during that time, though he would later return and fight as an officer under General Bonaparte until resigning his position due to ill health. He returned home and there, helped by his family wealth, he spent the rest of his life as an inventor. It was at this time that he and his brother built an internal combustion engine and used it to power a boat.[51] However, it was in the field of photography that he would do his most important work.

Lithography is a method of transferring an image onto a metal sheet or a flat stone. It was a very popular hobby in France in the early 1800s. Niépce's dabbling with lithographs led to a search to find a chemical coating that could be used to permanently capture an image. His experimentation included a wide variety of substances—silver chloride, silver nitrate, bitumen, lycopodium powder, white kerosene, and various acids and solutions. His searching was probably not unlike Edison's search a hundred years later for just the right filament for his light bulb.

Niépce finally succeeded using Bitumen of Judea. It was a substance which hardened in the brightly lit areas but could be washed away from dimly lit areas. He pointed his camera out a window at his country estate, Le Gras, and his *View from the Window at Le Gras* was the result. Due to its long exposure time, eight hours, its changing shadows through the day caused a less than sharp image, but it was the beginning of photography. He called it heliography—sun drawing. (The word "photography" would not exist for another decade.) He would make improvements in the process but was not able to appreciably reduce the exposure time, making portraits of people or scenes with any movement impractical.

He formed a partnership with Louis Daguerre in 1829. Daguerre knew of Niépce's work and acknowledged his priority in making the first photograph. He felt their combined efforts would prove beneficial, which indeed was the case.

Daguerre was a stage designer and painter of scenes for the Paris Opera. His "daguerreotypes" would be the first form of popular photography. He was able to decrease the exposure time from the restrictive eight hours to under one minute. Niépce's photos were made from an original negative image, with light areas being dark and dark areas being white. The daguerreotypes were a positive image. Niépce would not share in much of the partnership, passing away after only four years. However, in 1839, Niépce's son, Isadore, along with Daguerre, would be given a substantial monetary award by the French government in exchange for rights to their process.

Afterword

By 1815, the United States had elected four presidents. There were some acts of insurrection such as the Whiskey Rebellion and Shay's Rebellion, but these resulted in only a handful of deaths and most arrests resulted in amnesty. There were two constitutions—the Articles of Confederation and the current United States Constitution. There were no coups. There was one basic form of government—a republic.

John Adams became the second president. At his inauguration, President Washington and he marched into Congress Hall in Philadelphia. Washington sat down and viewed John Adams being sworn in. The ceremony complete, they departed. John Adams got to work, and George Washington went home to Mount Vernon.[1] This took place in 1797. At the same time, France was at war at home and abroad. France had been in a state of upheaval for a decade and would continue to be for many more years.

The amount of French scientific discovery during this time is remarkable. The French Revolution, quickly followed by the reign of Napoléon Bonaparte were times of chaos. There certainly have been other times of upheaval—the fall of the Roman Empire, the American Civil War, the Russian Revolution. Yet those times did not produce nearly the scientific progress that occurred in France. In fact, quite the opposite. Progress was put on hold, waiting for a more peaceful time.

The discoveries in France were probably rivaled only by those in the United States a hundred years later. In a comparable amount of time, Alexander Graham Bell's telephone (1876); the Wright Brothers' airplane (1903); along with Thomas Edison's electric light bulb (1879), phonograph (1887), and moving pictures (late 1880s) were invented. Those inventions, and Edison and Nikola Tesla's development of direct and alternating current, certainly changed the world. However, the Gilded Age was a very pro-business time in the United States, and except for the four-month Spanish-American War, it was a time of peace for the country. It was quite a different situation than what was taking place in revolutionary France.

So, what are the answers to the original issues posed? How could this have been both the best and the worst of times? How could such creativity and advancement come about during a time and place that were in complete turmoil?

The Enlightenment certainly supported a love of learning and growth that could motivate people to create. Perhaps another reason was simply that good fortune fell France's way. Those with ability and drive such as Antoine Lavoisier, Napoléon Bonaparte, and Jean Champollion do not come along often.

Could the chaos of the time have caused the creativity? It seems doubtful, but perhaps. If so, nothing like it existed in world history. More likely, it speaks to the human spirit that these advances took place not because of that chaos, but in spite of it. Perhaps it gives us hope that even in the darkest times, order can triumph over disorder, sanity over insanity, and good intentions over bad.

Chapter Notes

Introduction

1. Jackson, *A World on Fire*, p. 282.

Chapter I

1. Hardman, *The Life of Louis XVI*, p. 438.
2. Unger, *Thomas Paine and the Clarion Call for American Independence*, p. 171.
3. Hardman, p. 438.
4. Hibbert, *The French Revolution*, p. 186.
5. Hibbert, p. 186.
6. Hardman, p. 442.
7. Fraser, *Marie Antoinette: The Journey*, p. 445.
8. Hibbert, p. 222.
9. Miller and Molesky, *Our Oldest Enemy*, p. 41.
10. Unger, p. 192.
11. Philbrick, *In the Hurricane's Eye*, p. 276.
12. Fraser, p. 248.
13. Summers, *A Biography of Nicolas Appert 1749–1841*, p. 38.
14. Hibbert, p. 59.
15. Alder, *Engineering the Revolution*, p. 3.
16. Cooper, *Talleyrand*, p. 41.
17. Fraser, p. 297.
18. Chernow, *Alexander Hamilton*, p. 433.
19. Auricchio, *The Marquis: Lafayette Reconsidered*, p. 280.
20. Horne, *La Belle France: A Short History*, p. 197.
21. Moore, *Hung, Drawn, and Quartered: The Story of Execution Through the Ages*, p. 159.
22. "Reign of Terror," *Encyclopedia Britannica*, 29 August 2022, www.britannica.com/event/Reign-of-Terror.

Chapter II

1. Isaacson, *Leonardo da Vinci*, p. 181.
2. Rolt, *The Balloonists: The History of the First Aeronauts*, p. 27.
3. Gillispie, *The Montgolfier Brothers and the Invention of Aviation*, p. 17.
4. Gillispie, *The Montgolfier Brothers and the Invention of Aviation*, p. 17.
5. Gillispie, *The Montgolfier Brothers and the Invention of Aviation*, p. 21.
6. Gillispie, *The Montgolfier Brothers and the Invention of Aviation*, p. 21.
7. Christopher, *Riding the Jetstream*, p. 5.
8. Gillispie, *The Montgolfier Brothers and the Invention of Aviation*, p. 3.
9. Rolt, p. 30.
10. Jackson, D., *The Aeronauts*, p. 12.
11. Dwiggins, *Riders of the Winds*, p. 17.
12. Rolt, p. 33.
13. Gillispie, *The Montgolfier Brothers and the Invention of Aviation*, p. 32.
14. Jackson, D., p. 14.
15. Jackson, D., p. 15.
16. Rolt, p. 36.
17. Gillispie, *The Montgolfier Brothers and the Invention of Aviation*, p. 18.
18. Gillispie, *The Montgolfier Brothers and the Invention of Aviation*, p. 22.
19. "Argand burner," *Encyclopedia Britannica*, 25 April 2013, www.britannica.com/technology/Argand-burner.
20. Jackson, D. p. 16.
21. Gillispie, *The Montgolfier Brothers and the Invention of Aviation*, p. 39.
22. Gillispie, *The Montgolfier Brothers and the Invention of Aviation*, p. 40.

23. Gillispie, *The Montgolfier Brothers and the Invention of Aviation*, p. 43.

24. Rolt, p. 40.

25. Gillispie, *The Montgolfier Brothers and the Invention of Aviation*, p. 42.

26. Rolt, p. 42.

27. Rolt, p. 42.

28. Rolt, p. 37.

29. Dwiggins, p. 22.

30. Gillispie, *The Montgolfier Brothers and the Invention of Aviation*, pp. 47, 48.

31. Dwiggins, p. 22.

32. Rolt, p. 46.

33. Jackson, D. p. 20.

34. Simons, *The Early History of Ballooning*, p. 57.

35. Jackson, D. p. 18.

36. Christopher, p. 10.

37. Jackson, D., p. 24.

38. Jackson, D., p. 24.

39. Rolt, p. 53.

40. Christopher, p. 10.

41. Christopher, p. 12.

42. Gillispie, *The Montgolfier Brothers and the Invention of Aviation*, p. 62.

43. Jackson, D., p. 24.

44. Jackson, D., p. 27.

45. Jackson, D., p. 27.

46. Jackson, D., p. 27.

47. Howard, *Wilbur and Orville: A Biography of the Wright Brothers*, pp. 223, 224.

48. Rolt, p. 55.

49. Gillispie, *The Montgolfier Brothers and the Invention of Aviation*, p. 135.

50. Holmes, *Falling Upwards: How We Took to the Air*, p. 164.

51. Holmes, *Falling Upwards: How We Took to the Air*, p. 165.

52. Rolt, p. 148.

53. Jackson, D., p. 64.

54. Rolt, p. 152.

55. Rolt, p. 204.

56. Abrams, *John Adams Under Fire*, p. 203.

57. Rolt, p. 86.

58. Rolt, p. 87.

59. Rolt, p. 88.

60. Rolt, p. 91.

61. Jackson, D., p. 35.

62. Dwiggins, p. 36.

63. Gillispie, *The Montgolfier Brothers and the Invention of Aviation*, p. 87.

64. Dwiggins, p. 32.

65. Rolt, p. 89.

66. Jackson, D., p. 47.

67. Holmes, *Falling Upwards: How We Took to the Air*, p. 43.

68. Jackson, D., p. 52.

69. Holmes, *Falling Upwards: How We Took to the Air*, p. 39.

70. Holmes, *Falling Upwards: How We Took to the Air*, p. 35.

71. Jackson, D., p. 49.

72. Rolt, p. 161.

73. Jackson, D., p. 80.

74. Holmes, *Falling Upwards: How We Took to the Air*, p. 126.

75. Holmes, *Falling Upwards: How We Took to the Air*, p. 127.

76. Holmes, *Falling Upwards: How We Took to the Air*, p. 142.

77. Dwiggins, p. 114.

78. Christopher, p. 50.

79. Jackson, D., p. 111.

80. Jackson, D., p. 112.

81. McCullough, *Wright Brothers*, p. 1.

Chapter III

1. Bell, M., *Lavoisier in the Year One: The Birth of a New Science in an Age of Revolution*, p. 62.

2. Holmes, *The Age of Wonder*, p. 245.

3. Grey, *The Chemist Who Lost His Head*, p. 73.

4. Jackson, J., *A World on Fire: A Heretic, an Aristocrat, and the Race to Discover Oxygen*, p. 76.

5. Kjelle, *Antoine Lavoisier*, p. 16.

6. Bell, M., p. 19.

7. Bell, M., p. 14.

8. Biography.com Editors. "Jacques Louis David." *Biography*, April 2, 2014, www.biography.com/artist/jacques-louis-david.

9. Strathern, *Mendeleyev's Dream: The Quest for the Elements*, p. 37.

10. Strathern, *Mendeleyev's Dream: The Quest for the Elements*, p. 66.

11. Strathern, *Mendeleyev's Dream: The Quest for the Elements*, p. 17.

12. Strathern, *Mendeleyev's Dream: The Quest for the Elements*, p. 21.

13. Strathern, *Mendeleyev's Dream: The Quest for the Elements*, p. 185.

14. Strathern, *Mendeleyev's Dream: The Quest for the Elements*, p. 184.

15. Strathern, *Mendeleyev's Dream: The Quest for the Elements*, p. 208.

16. Kjelle, p. 24.

17. Strathern, *Mendeleyev's Dream: The Quest for the Elements*, p. 218.
18. Strathern, *Mendeleyev's Dream: The Quest for the Elements*, p. 197.
19. Kjelle, p. 26.
20. Jackson, J., pp. 65, 66.
21. Johnson, *The Invention of Air*, p. xiv.
22. Strathern, *Mendeleyev's Dream: The Quest for the Elements*, p. 223.
23. Bell, M., p. 85.
24. Jackson, J., p. 70.
25. Jackson, J., p. 202.
26. Jackson, J., p. 255.
27. Poirier, *Lavoisier: Chemist, Biologist, Economist*, p. 11.
28. Bell, M., p. 90.
29. Bell, M., p. 91.
30. Bell, M., p. 62.
31. Poirier, p. 51.
32. Jackson, J., p. 215.
33. Poirier, p. 63.
34. Bell, M., p. 91.
35. Bell, M., p. 111.
36. Bell, M., p. 110.
37. Susac, *The Clock, the Balance, and the Guillotine: The Life of Antoine Lavoisier*, p. 4.
38. Strathern, *Mendeleyev's Dream: The Quest for the Elements*, p. 233.
39. Poirier, p. 143.
40. Lavoisier, *Elements of Chemistry*, p. 1.
41. Jackson, J., p. 218.
42. Bell, M., p. 138.
43. Bell, M., pp. 137, 138.
44. Poirier, p. 187.
45. Jackson, J., p. 222.
46. Poirier, p. 121.
47. Kjelle, Marylou Morano, p. 32.
48. Poirier, p. 129.
49. Poirier, p. 223.
50. Bell, M., p. 22.
51. Kjelle, p. 27.
52. Bell, M., p. 92.
53. Bell, M., p. 24.
54. Bell, M., p. 31.
55. Bell, M., p. 31.
56. Holmes, *The Age of Wonder*, p. 314.
57. Jackson, J., p. 274.
58. Jackson, J., p. 278.
59. Jackson, J., p. 286.
60. Jackson, J., p. 286.
61. Susac, p. 11.
62. Jackson, J., p. 93.
63. Hibbert, p. 222.
64. Jackson, J., p. 92.
65. Bell, M., p. 16.
66. Susac, p. 34.
67. Kjelle, p. 33.
68. Bell, M., p. 27.
69. Bell, M., p. 175.
70. Poirer, p. 390.
71. Poirier, p. 393.
72. Poirier, p. 371.
73. Poirier, p. 379.
74. Poirier, p. 378.
75. Poirier, p. 379.
76. Jackson, J., p. 327.
77. Poirier, p. 461.
78. Kjelle, p. 40.
79. Poirier, p. 386.
80. Bell, M., p. 173.
81. Poirier, p. 385.
82. Poirier, p. 404.

Chapter IV

1. Crease, *World in the Balance*, p. 102.
2. Robinson, *The Story of Measurement*, p. 27.
3. Alder, *The Measure of All Things*, p. 95.
4. Marciano, *Whatever Happened to the Metric System?*, p. 22.
5. Marciano, p. 48.
6. Crease, p. 88.
7. Lawday, *Napoleon's Master: A Life of Prince Talleyrand*, p. 98.
8. Cooper, *Talleyrand*, p. 28.
9. Cooper, p. 176.
10. Alder, *The Measure of All Things*, p. 91.
11. Alder, *The Measure of All Things*, p. 89.
12. Marciano, p. 73.
13. Crease, p. 91.
14. Robinson, *The Story of Measurement*, p. 25.
15. Crease, p. 86.
16. Alder, *The Measure of All Things*, p. 96.
17. Crease, p. 91.
18. Alder, *The Measure of All Things*, p. 156.
19. Alder, *The Measure of All Things*, p. 45.
20. Alder, *The Measure of All Things*, p. 55.
21. Alder, *The Measure of All Things*, p. 154.
22. Alder, *The Measure of All Things*, p. 227.

23. Alder, *The Measure of All Things*, p. 209.
24. Alder, *The Measure of All Things*, p. 70.
25. Alder, *The Measure of All Things*, pp. 111, 112.
26. Alder, *The Measure of All Things*, p. 240.
27. Alder, *The Measure of All Things*, p. 230.
28. Alder, *The Measure of All Things*, p. 235–237.
29. Alder, *The Measure of All Things*, p. 7.
30. Alder, *The Measure of All Things*, p. 264.
31. Alder, *The Measure of All Things*, p. 356.
32. Alder, *The Measure of All Things*, p. 227.
33. Marciano, p. 95.
34. Marciano, p. 262.
35. Crease, p. 101.
36. Alder, *The Measure of All Things*, p. 161.
37. Marciano, p. 71.
38. Robinson, *The Story of Measurement*, p. 29.

Chapter V

1. Roberts, *Napoleon: A Life*, p. 732.
2. Roberts, p. 317.
3. Schom, *Napoleon Bonaparte*, p. 103.
4. Schom, p. 187.
5. Roberts, p. 75.
6. Zamoyski, *Napoleon: A Life*, p. 336.
7. Roberts, p. 201.
8. Roberts, p. 11.
9. Roberts, p. 50.
10. Roberts, p. 66.
11. Cooper, *Talleyrand*, p. 107.
12. Price, *Napoleon: The End of Glory*, p. 61.
13. Roberts, p. 239.
14. Meyerson, *The Linguist and the Emperor*, p. 148.
15. Arana, *Bolívar: American Liberator*, p. 59.
16. Holtman, *The Napoleonic Revolution*, p. 87.
17. Holtman, p. 88.
18. Horne, *La Belle France*, p. 213.
19. Horne, p. 210.
20. Roberts, p. 689.

21. Zamoyski, p. 385.
22. Zamoyski, p. 254.
23. Adkins, *Nelson's Trafalgar*, p. 279.
24. Schom, p. 137.
25. Roberts, p. 278.
26. Zamoyski, p. 522.
27. Zamoyski, p. 562.
28. Price, p. 58.
29. Price, p. 5.
30. Price, p. 34.
31. Price, p. 144.

Chapter VI

1. Shaw, *Ancient Egypt: A Very Short Introduction*, p. 66.
2. Roberts, p. 158.
3. Burleigh, *Mirage: Napoleon's Scientists and the Unveiling of Egypt*, p. 37.
4. Strathern, *Napoleon in Egypt*, p. 236.
5. Brier, *Egyptomania: Our Three Thousand Year Obsession with the Land of the Pharaohs*, p. 45.
6. Alder, *The Measure of All Things*, p. 254.
7. Brier, p. 60.
8. Dwyer, *Napoleon: The Path to Power*, p. 359.
9. Dwyer, p. 364.
10. Schom, p. 124.
11. Schom, p. 116.
12. Strathern, *Napoleon in Egypt*, p. 86.
13. Strathern, *Napoleon in Egypt*, p. 120.
14. Roberts, p. 166.
15. Roberts, p. 175.
16. Roberts, p. 272.
17. Strathern, *Napoleon in Egypt*, p. 140.
18. Strathern, *Napoleon in Egypt*, p. 250.
19. Roberts, p. 174.
20. Roberts, p. 799.
21. Dwyer, p. 383.
22. Dwyer, pp. 403, 404.
23. Roberts, p. 192.
24. Burleigh, p. 96.
25. Burleigh, p. 146.
26. Burleigh, p. 146.
27. Strathern, *Napoleon in Egypt*, p. 254.
28. Inglis-Arkell, "The Scientist Who Inspired the Count of Monte Cristo," *Gizmodo*, June 9, 2014, www.gizmodo.com/the-scientist-who-inspired-the-count-of-monte-cristo-1588185366.
29. Schom, p. 142.
30. Schom, p. 139.
31. Schom, p. 143.

32. Strathern, *Napoleon in Egypt*, p. 147.
33. Alder, *The Measure of All Things*, p. 187.
34. Burleigh, p. 230.
35. Brier, p. 60.
36. Burleigh, p. 49.
37. Brier, p. 60.
38. Burleigh, p. 104.
39. Burleigh, p. 112.
40. Roberts, p. 84.
41. Russell, *The Discovery of Egypt: Vivant Denon's Travels with Napoleon's Army*, p. 71.
42. Russell, *The Discovery of Egypt: Vivant Denon's Travels with Napoleon's Army*, p. 52.
43. Strathern, *Napoleon in Egypt*, p. 295.
44. Russell, *The Discovery of Egypt: Vivant Denon's Travels with Napoleon's Army*, p. 225.
45. Russell, *The Discovery of Egypt: Vivant Denon's Travels with Napoleon's Army*, p. 233.
46. Burleigh, p. 123.
47. Roberts, p. 189.
48. Burleigh, p. 118.
49. Strathern, *Napoleon in Egypt*, p. 310.
50. Strathern, *Napoleon in Egypt*, p. 98.
51. Strathern, *Napoleon in Egypt*, p. 373.
52. Schom, p. 179.
53. Schom, p. 179.
54. Adkins, *The Keys of Egypt*, p. 38.

Chapter VII

1. Shaw, *Ancient Egypt: A Very Short Introduction*, p. 104.
2. Brier, p. 32.
3. Ray, *The Rosetta Stone and the Rebirth of Ancient Egypt*, p. 72.
4. Ray, p. 4.
5. Strathern, *Napoleon in Egypt*, p. 389.
6. Robinson, *Cracking the Egyptian Code*, p. 88.
7. Giblin, *The Riddle of the Rosetta Stone*, p. 11.
8. Robinson, *Cracking the Egyptian Code*, p. 16.
9. Buchwald, *The Riddle of the Rosetta*, p. 23.
10. Robinson, *Cracking the Egyptian Code*, p. 81.
11. Robinson, *Cracking the Egyptian Code*, p. 81.
12. Ray, p. 42.
13. Buchwald, p. 280.
14. Buchwald, p. 297.
15. Meyerson, *The Linguist and the Emperor*, p. 174.
16. Adkins, *The Keys of Egypt: The Obsession to Decipher Egyptian Hieroglyphs*, p. 71.
17. Adkins, *The Keys of Egypt: The Obsession to Decipher Egyptian Hieroglyphs*, p. 123.
18. Robinson, *Cracking the Egyptian Code*, p. 94.
19. Meyerson, p. 220.
20. Robinson, *Cracking the Egyptian Code*, p. 92.
21. Robinson, *Cracking the Egyptian Code*, p. 106.
22. Robinson, *Cracking the Egyptian Code*, p. 126.
23. Robinson, *Cracking the Egyptian Code*, p. 93.
24. Robinson, *Cracking the Egyptian Code*, p. 139.
25. Giblin, p. 32.
26. Adkins, *The Keys of Egypt: The Obsession to Decipher Egyptian Hieroglyphs*, p. 64.
27. Robinson, *Cracking the Egyptian Code*, p. 79.
28. Adkins, *The Keys of Egypt: The Obsession to Decipher Egyptian Hieroglyphs*, p. 157.
29. Buchwald, p. 272.
30. Robinson, *Cracking the Egyptian Code*, p. 88.
31. Robinson, *Cracking the Egyptian Code*, p. 89.
32. Ray, pp. 49, 50.
33. Robinson, *Cracking the Egyptian Code*, p. 123.
34. Robinson, *Cracking the Egyptian Code*, p. 149.
35. Buchwald, p. 389.
36. Buchwald, p. 82.
37. Adkins, *The Keys of Egypt: The Obsession to Decipher Egyptian Hieroglyphs*, p. 247.
38. Robinson, *Cracking the Egyptian Code*, p. 200.
39. Robinson, *Cracking the Egyptian Code*, p. 205.
40. Robinson, *Cracking the Egyptian Code*, p. 235.
41. Giblin, p. 64.

Chapter VIII

1. Wilson, *Charles Darwin: Victorian Mythmaker*, p. 26.
2. Wilson, p. 28.
3. Stott, *Darwin's Ghosts: The Secret History of Evolution*, p. 175.
4. Stott, p. 179.
5. Stott, p. 37.
6. Wagner, *Arrival of the Fittest: Solving Evolution's Greatest Puzzle*, p. 8.
7. Reis, *Not by Design: Retiring Darwin's Watchmaker*, p. 89.
8. Prothero, *The Story of Evolution in 25 Discoveries*, p. 289.
9. Prothero, *The Story of Evolution in 25 Discoveries*, p. 212.
10. Morgan, *Benjamin Franklin*, p. 8.
11. Stott, p. 193.
12. Stott, p. 196.
13. Prothero, *The Story of Evolution in 25 Discoveries*, p. 212.
14. Stott, p. 204.
15. Prothero, *The Story of Evolution in 25 Discoveries*, p. 198.
16. Wilson, *Charles Darwin: Victorian Mythmaker*, p. 17.
17. Sawan and Parvizi, "Science of Epigenetics: Lamarck was right after all," *Healio*, June 01, 2013, www.healio.com/news/orthopedics/20130611/10_3928_1081_597x_20130101_02_1167467.
18. Parissien, *The Life of the Automobile*, p. 4.
19. Parissien, p. 4.
20. Caullery, *French Science and Its Principal Discoveries since the Seventeenth Century*, p. 62.
21. McCloy, *French Inventions of the Eighteenth Century*, p. 32.
22. McCloy, p. 33.
23. McCloy, p. 36.
24. McCloy, pp. 58, 59.
25. Alder, *Engineering the Revolution: Arms and Enlightenment in France 1763–1815*, p. 3.
26. Harford, "How Interchangeable Parts Revolutionized the Way Things Are Made," *BBC*, 9 October 2019, www.bbc.com/news/business-49499444.
27. Summers, *A Biography of Nicolas Appert 1749–1841*, p. 68.
28. Appert, p. 135.
29. Appert, p. 135.
30. Appert, p. 7.
31. Snodgrass, *Encyclopedia of Kitchen History*, p. 29.
32. Summers, p. 142.
33. Solymar, *Getting the Message: A History of Communications*, p. 26.
34. Robinson, *The Story of Measurement*, p. 62.
35. Robinson, *The Story of Measurement*, p. 62.
36. Solymar, p. 33.
37. Solymar, p. 30.
38. Essinger, *Jacquard's Web: How a Hand-Loom Led to the Birth of the Information Age*, p. 24.
39. Essinger, *Jacquard's Web: How a Hand-Loom Led to the Birth of the Information Age*, p. 25.
40. Essinger, *Ada's Algorithm*, p. 76.
41. Essinger, *Jacquard's Web: How a Hand-Loom Led to the Birth of the Information Age*, p. 280.
42. Essinger, *Jacquard's Web: How a Hand-Loom Led to the Birth of the Information Age*, p. 39.
43. Essinger, *Jacquard's Web: How a Hand-Loom Led to the Birth of the Information Age*, p. 47.
44. Essinger, *Jacquard's Web: How a Hand-Loom Led to the Birth of the Information Age*, p. 141.
45. Essinger, *Jacquard's Web: How a Hand-Loom Led to the Birth of the Information Age*, p. 174.
46. Rhodes, *Hedy's Folly*, p. 147.
47. Rhodes, p. 172.
48. Cheslak, "Hedy Lamarr," *National Women's History Museum*, 30 August 2018, www.womenshistory.org/students-and-educators/biographies/hedy-lamarr.
49. Vallencourt, *The History of Photography*, p. 4.
50. Newhall, *The History of Photography: From 1839 to the present*, p. 13.
51. Newhall, p. 13.

Afterword

1. McCullough, *John Adams*, p. 467.

Bibliography

Abrams, Dan, and David Fisher. *John Adams Under Fire: The Founding Father's Fight for Justice in the Boston Massacre Murder Trial*. Toronto: Hanover Square Press, 2021.

Adkins, Lesley, and Roy A. Adkins. *The Keys of Egypt: The Obsession to Decipher Egyptian Hieroglyphs*. New York: HarperCollins, 2000.

Adkins, Roy. *Nelson's Trafalgar: The Battle That Changed the World*. New York: Viking, 2005.

———, and Lesley Adkins. *The War for All the Oceans: From Nelson at the Nile to Napoleon at Waterloo*. New York: Penguin Books, 2008.

Alder, Ken. *Engineering the Revolution: Arms and Enlightenment in France 1763–1815*. Chicago, University of Chicago Press, 2010.

———. *The Measure of All Things: The Seven-Year Odyssey That Transformed the World*. London, Little, Brown, 2002.

Anderson, Fred. *The War That Made America: A Short History of the French and Indian War*. New York: Penguin Books, 2006.

Appert, Nicolas. *Preserving All Kinds of Vegetables for Several Years*. London: Black, Parry, and Kingsbury, 1812.

Arana, Marie. *Bolívar: American Liberator*. New York: Simon & Schuster Paperbacks, 2014.

"Argand Burner." *Encyclopedia Britannica*, 25 April 2013, https://www.britannica.com/technology/Argand-burner. Accessed 26 August 2022.

Auricchio, Laura. *The Marquis: Lafayette Reconsidered*. Novato: Presidio, 2015.

Bell, David. *Napoleon. A Concise Biography*. Oxford: Oxford University Press, 2016.

Bell, Madison Smartt. *Lavoisier in the Year One: The Birth of a New Science in an Age of Revolution*. New York: W.W. Norton, 2005.

Bell, T. F. *Jacquard Weaving and Designing*. United States: Createspace Independent Publishing Platform, 2017.

Biography.com Editors. "Jacques Louis David." *Biography*, April 2, 2014, www.biography.com/artist/jacques-louis-david.

Boorstin, Daniel J. *Seekers the Story of Man's Continuing Quest to Understand His World*. New York: Random House, 2000.

Boyer, Carl Benjamin, and Uta C. Merzbach. *A History of Mathematics*. Hoboken, NJ: Wiley, 1989.

Brier, Bob. *Egyptomania: Our Three Thousand Year Obsession with the Land of the Pharaohs*. New York: Palgrave Macmillan, 2013.

Broers, Michael. *Napoleon. Soldier of Destiny*. New York: Pegasus Books, 2015.

Buchwald, Jed, and Diane Greco Josefowicz. *The Riddle of the Rosetta: How an English Polymath and a French Polyglot Discovered the Meaning of Egyptian Hieroglyphs*. Princeton: Princeton University Press, 2020.

Burleigh, Nina. *Mirage: Napoleon's Scientists and the Unveiling of Egypt*. New York: Harper/Perennial, 2008.

Caullery, Maurice. *French Science and Its Principal Discoveries Since the Seventeenth Century*. New York: Arno Press, 1975.

Chaplin, Joyce E. *The First Scientific American: Benjamin Franklin and the Pursuit of Genius*. New York: Basic Books, 2007.

Chernow, Ron. *Alexander Hamilton*. New York: Penguin Books, 2005.

Cheslak, Colleen. "Hedy Lamarr." *Hedy Lamarr*, National Women's History Museum, 30 August 2018, www.womenshistory.org/students-and-educators/biographies/hedy-lamarr.

Christopher, John. *Riding the Jetstream: The Story of Ballooning: From Montgolfier to Bretling*. London: John Murray, 2001.

Cobb, Richard, and Colin Jones, eds. *The French Revolution: Voices from a Momentous Epoch, 1789–1795*. London: Simon & Schuster, 1988.

Cole, Juan Ricardo. *Napoleon's Egypt: Invading the Middle East*. New York: St. Martin's Griffin, 2008.

Connolly, Owen. *The Wars of the French Revolution and Napoleon, 1792–1815*. London: Routledge, 2005.

Cooper, Duff. *Talleyrand*. London: Phoenix Giant, 1997.

Crease, Robert P. *World in the Balance: The Historic Quest for an Absolute System of Measurement*. New York: W.W. Norton, 2012.

Darwin, Charles, et al. *Charles Darwin: Autobiographies*. London: Penguin, 2002.

Durant, Will, and Ariel Durant. *The Age of Napoleon: A History of European Civilization from 1789 to 1815*. New York: Simon & Schuster, 1975.

———. *The Age of Voltaire: A History of Civilization in Western Europe from 1715 to 1756. With Special Emphasis on the Conflict Between Religion and Philosophy*. New York: Simon & Schuster, 1965.

———. *Rousseau and Revolution: A History of Civilization in France, England, and Germany from 1756, and in the Remainder of Europe from 1715 to 1789*. New York: MJF Books, 1992.

Dwiggins, Don. *Riders of the Winds: The Story of Ballooning*. New York: Hawthorn Books, 1973.

Dwyer, Philip G. *Napoleon: The Path to Power*. New Haven, CT: Yale University Press, 2008.

Erickson, Carolly. *Josephine: A Life of the Empress*. New York: St. Martin's Press, 2000.

Essinger, James. *Ada's Algorithm: How Lord Byron's Daughter Ada Lovelace Launched the Digital Age*. Brooklyn, NY: Melville House, 2015.

———. *Jacquard's Web: How a Hand-Loom Led to the Birth of the Information Age*. Oxford, UK: Oxford University Press, 2004.

Farndon, John. *The Great Scientists: From Euclid to Stephen Hawking*. London: Hinkler Books, 2010.

Ferreiro, Larrie D. *Measure of the Earth: The Enlightenment Expedition That Reshaped the World*. New York: Basic Books, 2013.

Fraser, Antonia. *Marie Antoinette: The Journey*. New York: Anchor Books, 2002.

Giblin, James. *The Riddle of the Rosetta Stone: Key to Ancient Egypt: Illustrated with Photographs, Prints, and Drawings*. New York: HarperTrophy, 1992.

Gillispie, Charles Coulston. *The Montgolfier Brothers and the Invention of Aviation, 1783–1784 with a Word on the Importance of Ballooning for the Science of Heat and the Art of Building Railroads*. Princeton, NJ: Princeton University Press, 2014.

———. *Pierre-Simon Laplace, 1749–1827: A Life in Exact Science*. Princeton, NJ: Princeton University Press, 1997.

———. *Science and Polity in France: The Revolutionary and Napoleonic Years*. Princeton, NJ: Princeton University Press, 2004.

Grew, Raymond. *Food in Global History*. Boulder, CO: Westview Press, 1999.

Grey, Vivien. *The Chemist Who Lost His Head: Antoine Laurent Lavoisier*. New York: Coward, McCann & Geoghegan, 1982.

Hankins, Thomas L. *Science and the Enlightenment*. Cambridge: Cambridge University Press, 1985.

Hardman, John. *The Life of Louis XVI*. New Haven, CT: Yale University Press, 2016.

Harris, Robin. *Talleyrand: Betrayer and Saviour of France*. London: John Murray, 2007.

Hibbert, Christopher. *The French Revolution*. London: Penguin, 1982.

Holmes, Richard. *The Age of Wonder: How the Romantic Generation Discovered the Beauty and Terror of Science*. New York: Pantheon Books, 2008.

———. *Falling Upwards: How We Took to the Air*. New York: Pantheon Books, 2013.

Holtman, Robert B. *The Napoleonic Revolution*. Philadelphia: Lippincott, 1995.

Horne, Alistair. *La Belle France: A Short History.* New York: Vintage Books, 2006.

Howard, Fred. *Wilbur and Orville: A Biography of the Wright Brothers.* New York: Knopf, 1988.

Inglis-Arkell, Esther. "The Scientist Who Inspired the Count of Monte Cristo." *Gizmodo,* June 9, 2014, www.gizmodo.com/the-scientist-who-inspired-the-count-of-monte-cristo-1588185366.

Isaacson, Walter. *Leonardo da Vinci.* New York: Simon & Schuster, 2017.

Jackson, Donald Dale, and Time-Life Books. *The Aeronauts.* Alexandria, VA: Time-Life Books, 1981.

Jackson, Joe. *A World on Fire: A Heretic, an Aristocrat, and the Race to Discover Oxygen.* New York: Penguin Books, 2007.

Johnson, Steven. *The Invention of Air: A Story of Science, Faith, Revolution, and the Birth of America.* New York: Riverhead Books, 2008.

Jones, Colin. *Voices of the French Revolution.* Topsfield, MA: Salem House Publishers, 1988.

Kean, Sam. *Caesar's Last Breath: The Epic Story of the Air Around Us.* London: Doubleday, 2017.

Kjelle, Marylou Morano. *Antoine Lavoisier: Father of Modern Chemistry.* Hockessin, DW: Mitchell Lane Publishers, 2005.

Kotar, S. L., and J. E. Gessler. *Ballooning: A History, 1782–1900.* Jefferson, NC: McFarland, 2011.

Lavoisier, Antoine Laurent. *Elements of Chemistry, in a New Systematic Order, Containing All the Modern Discoveries.* New York: Dover Publications, 1965.

Lawday, David. *Napoleon's Master: A Life of Prince Talleyrand.* New York: Thomas Dunne Books/St. Martin's Press, 2007.

Levy, Joel. *History's Greatest Discoveries and the People Who Made Them.* New York: Metro Books, 2015.

Littlewood, Ian, and Justin Wintle. *The Timeline History of France.* New York: Barnes & Noble, 2005.

Mannion, James. *Essential Philosophy: Everything You Need to Understand the World's Greatest Thinkers.* Avon, MA: Adams Media Corp, 2006.

Marciano, John Bemelmans. *Whatever Happened to the Metric System? How America Kept Its Feet.* New York: Bloomsbury, 2015.

Maurice, Felix. *Napoleon.* New York: Signet Classics, 2010.

McCloy, Shelby T. *French Inventions of the Eighteenth Century.* Lexington: University of Kentucky Press, 2014.

McCullough, David. *John Adams.* New York: Simon & Schuster Paperbacks, 2004.

_____. *Wright Brothers.* New York: Simon & Schuster, 2015.

Meyerson, Daniel. *The Linguist and the Emperor: Napoleon and Champollion's Quest to Decipher the Rosetta Stone.* New York: Random House Trade Paperbacks, 2005.

Miller, John J., and Mark Molesky. *Our Oldest Enemy: A History of America's Disastrous Relationship with France.* New York: Doubleday, 2004.

Moore, Jonathan J. *Hung, Drawn, and Quartered: The Story of Execution Through the Ages.* New York: Metro Books, 2017.

Morgan, Edmund S. *Benjamin Franklin.* New Haven, CT: Yale University Press, 2002.

Newhall, Beaumont. *The History of Photography: From 1839 to the Present.* New York: Museum of Modern Art, 1982.

Parissien, Steven. *The Life of the Automobile.* New York: Thomas Dunne Books, 2014.

Philbrick, Nathaniel. *In the Hurricane's Eye: The Genius of George Washington and the Victory at Yorktown.* New York: Penguin Books, 2019.

Poirier, Jean-Pierre. *Lavoisier: Chemist, Biologist, Economist.* Philadelphia: University of Pennsylvania Press, 1996.

Posamentier, Alfred S. and Christian Spreitzer. *Math Makers: The Lives and Works of 50 Famous Mathematicians.* Amherst, NY: Prometheus Books, 2019.

Price, Munro. *Napoleon: The End of Glory.* Oxford, UK: Oxford University Press, 2014.

Prothero, Donald R. *The Story of Evolution in 25 Discoveries: The Evidence and the People Who Found It.* New York: Columbia University Press, 2020.

_____, and Carl Dennis Buell. *Evolution: What the Fossils Say and Why It Matters.* New York: Columbia University Press, 2017.

Ray, J. D. *The Rosetta Stone and the Rebirth of Ancient Egypt.* Cambridge, MA: Harvard University Press, 2007.

"Reign of Terror." *Encyclopedia Britannica,*

29 August 2022, https://www.britannica.com/event/Reign-of-Terror. Accessed 29 August 2022.

Reiss, John O. *Not by Design: Retiring Darwin's Watchmaker*. Berkeley: University of California Press, 2009.

Rhodes, Richard. *Hedy's Folly: The Life and Breakthrough Inventions of Hedy Lamarr, the Most Beautiful Woman in the World*. New York: Vintage Books, 2012.

Roberts, Andrew. *Napoleon: A Life*. New York: Viking, 2014.

Robinson, Andrew. *Cracking the Egyptian Code: The Revolutionary Life of Jean-François Champollion*. New York: Oxford University Press, 2012.

_____. *The Story of Measurement*. London: Thames & Hudson, 2007.

Rolt, L. T. C. *The Balloonists: The History of the First Aeronauts*. Stroud, UK: Sutton Pub., 2006.

Rudwick, Martin J. S. *Georges Cuvier, Fossil Bones, and Geological Catastrophes: New Translations & Interpretations of the Primary Texts*. Chicago: University of Chicago Press, 1997.

Russell, Terence M. *The Discovery of Egypt: Vivant Denon's Travels with Napoleon's Army*. Stroud, UK: Sutton, 2005.

_____. *The Napoleonic Survey of Egypt: Description de l'Egypte: The Monuments and Customs of Egypt Selected Engravings and Texts*. Farnham, UK: Ashgate, 2001.

Sawan, Hind, and Javad Parvizi. "Science of Epigenetics: Lamarck Was Right After All." *Healio*, June 1, 2013, www.healio.com/news/orthopedics/20130611/10_39 28_1081_597x_20130101_02_1167467.

Schama, Simon. *Citizens: A Chronicle of the French Revolution*. New York: Knopf, 1992.

Schom, Alan. *Napoleon Bonaparte*. New York: HarperCollins, 1997.

Shaw, Ian. *Ancient Egypt: A Very Short Introduction, 2nd Edition*. Oxford, UK: Oxford University Press, 2021.

Simmons, John. *The Scientific 100: A Ranking of the Most Influential Scientists, Past and Present*. New York: Fall River Press, 2009.

Simons, Fraser. *The Early History of Ballooning: "The Age of the Aeronaut."* Paris: Macha Press, 2014.

Smith, Sanderson M. *Agnesi to Zeno: Over*

100 Vignettes from the History of Math. Berkeley, CA: Key Curriculum Press, 1996.

Snodgrass, Mary Ellen. *Encyclopedia of Kitchen History*. New York: Fitzroy Dearborn, 2004.

Solymar, L. *Getting the Message: A History of Communications*. Oxford, UK: Oxford University Press, 2021.

Stewart, Ian. *Significant Figures: The Lives and Work of Great Mathematicians*. New York: Basic Books, 2017.

Stott, Rebecca. *Darwin's Ghosts: The Secret History of Evolution*. New York: Spiegel & Grau, 2013.

Strathern, Paul. *Mendeleyev's Dream: The Quest for the Elements*. New York: Pegasus Books, 2019.

_____. *Napoleon in Egypt*. New York: Bantam Dell, 2007.

Summers, Malcolm. *A Biography of Nicolas Appert 1749–1841*. Downs Way Publishing, 2015.

Susac, Andrew. *The Clock, the Balance, and the Guillotine: The Life of Antoine Lavoisier*. Garden City, NY: Doubleday, 1970.

Unger, Harlow G. *Thomas Paine and the Clarion Call for American Independence*. New York: Da Capo Press, 2019.

Vallencourt, Margaret. *The History of Photography*. New York: Britannica Educational Publishing in Association with Rosen Educational Services, 2016.

Wagner, Andreas. *Arrival of the Fittest: Solving Evolution's Greatest Puzzle*. New York: Penguin Group, 2014.

Whitelaw, Ian. *A Measure of All Things: The Story of Man and Measurement*. New York: St. Martin's Press, 2007.

Wilson, A.N. *Charles Darwin: Victorian Mythmaker*. New York: Harper, An Imprint of HarperCollins Publishers, 2017.

Winchester, Simon. *The Perfectionists: How Precision Engineers Created the Modern World*. New York: Harper, An Imprint of HarperCollins Publishers, 2018.

Windrow, Martin, and Francis K. Mason. *The World's Great Military Leaders: Two Hundred of the Most Significant Names in Land Warfare, from the 10th to the 20th Century*. New York: Gramercy Books, Random House, 2000.

Zamoyski, Adam. *Napoleon: A Life*. New York: Basic Books, 2018.

Index

Numbers in **bold italics** indicate pages with illustrations

199